Rump / Krist

# Laboratory Manual for the Examination of Water, Waste Water, and Soil

VCH

© VCH Verlagsgesellschaft mbH, D-6940 Weinheim (Federal Republic of Germany), 1988

Distribution
VCH Verlagsgesellschaft, P.O. Box 101161, D-6940 Weinheim (Federal Republic of Germany)
Switzerland: VCH Verlags-AG, P.O. Box, CH-4020 Basel (Switzerland)
Great Britain and Ireland: VCH Publishers (UK) Ltd., 8 Wellington Court, Wellington Street,
    Cambridge CB1 1HW (Great Britain)
USA und Canada: VCH Publishers, Suite 909, 220 East 23rd Street, New York NY 10010-4606 (USA)

ISBN 3-527-26973-8 (VCH Verlagsgesellschaft)        ISBN 0-89573-851-1 (VCH Publishers)

H.H. Rump / H. Krist

# Laboratory Manual for the Examination of Water, Waste Water, and Soil

VCH

Dr. Hans Hermann Rump
Institut Fresenius
D-6204 Taunusstein
Federal Republic of Germany

Dipl.-Ing. (FH) H. Krist
Deutsche Gesellschaft für Technische
Zusammenarbeit (GTZ) GmbH
Dag-Hammarskjöld-Weg 1
D-6236 Eschborn 1
Federal Republic of Germany

> This book was carefully produced. Nevertheless, authors, editors and publisher do not warrant the information contained therein to be free of errors. Readers are advised to keep in mind that statements, data, illustrations, procedural details or other items may inadvertently be inaccurate.

Published jointly by
VCH Verlagsgesellschaft, Weinheim (Federal Republic of Germany)
VCH Publishers, New York, NY (USA)

Editor: Deutsche Gesellschaft für Technische Zusammenarbeit (GTZ) GmbH
Editorial Director: Dr. Hans F. Ebel and Dr. Christina Dyllick-Brenzinger
Production Manager: Dipl.-Wirt.-Ing. (FH) Bernd Riedel
Printing: betz-druck gmbh, D-6100 Darmstadt
Bookbinding: Georg Kränkl, D-6148 Heppenheim

Library of Congress Card No. 88-39700
British Library Cataloguing in Publication Data
**Rump, H.H.**
Laboratory manual for the examination of
water, waste water and soil.
1. Natural resources : Water. Chemical
analysis. Laboratory techniques 2. Soils.
Properties. Measurement. Laboratory
techniques – Manuals
I. Title II. Krist, H. II. Laborhandbuch
für die Untersuchung von Wasser, Abwasser
und Boden. English
628.1'61'028
ISBN 3-527-26973-8

CIP-Kurztitelaufnahme der Deutschen Bibliothek:

**Rump, Hans Hermann:**
Laboratory manual for the examination of water, waste water
and soil / H.H. Rump ; H. Krist. [Ed.: Dt. Ges. für Techn.
Zusammenarbeit (GTZ) GmbH]. – Weinheim ; Basel ;
Cambridge ; New York : VCH, 1988
Dt. Ausg. u. d. T.: Rump, Hans Hermann: Laborhandbuch für die
Untersuchung von Wasser, Abwasser und Boden
ISBN 3-527-26973-8 (Weinheim ...) Gb.
ISBN 0-89573-851-1 (Cambridge ...) Gb.
NE: Krist, Helmut;

© VCH Verlagsgesellschaft, D-6940 Weinheim (Federal Republic of Germany), 1988
All rights reserved (including those of translation into other languages). No part of this book may be reproduced in any form – by photoprint, microfilm, or any other means – nor transmitted or translated into a machine language without written permission from the publishers. Registered names, trademarks, etc. used in this book, even when not specifically marked as such, are not to be considered unprotected by law.
Printed in the Federal Republic of Germany

# Foreword

The rising consumption of water for domestic, agricultural and industrial purposes is placing an increasing burden on nature's household. Apart from the problems associated with supplying sufficient quantities, an equally important difficulty involves the availability of water of acceptable quality.

Within the third phase of the "International Hydrological Program" (1984–1989) as well as the "Decade for Drinking Water Supply and Sanitary Measures" of the United Nations (1981–1990), questions regarding the availability and distribution of drinking and industrial water, and their disposal, have gained in importance, in particular for the third world countries.

Prerequisites for coordinated and appropriate action in these areas are the rapid access to the necessary information and the awareness of new procedures covering the multitude of practical requirements. The Deutsche Gesellschaft für Technische Zusammenarbeit-GmbH-(German Agency for Technical Cooperation)-GTZ has taken up the challenge of supplying information and further education in the sectors water supply and water disposal, by commissioning publications in priority areas of interest. These include the technology associated with drinking water and waste water, canalization, industrial water treatment and the practice of water analysis.

The present compilation will aid the specialist active in the fields of water evaluation and judgement of water quality. Well proven, uncomplicated analytical techniques are emphasized. Particular importance is placed on methods for quality control, sampling and evaluation of experimental results. The publication is intended especially for use in training courses and in further education as well as for personnel involved in development aid schemes.

The GTZ combines its gratitude to the authors with the hope of a widespread distribution of the "Laboratory Manual" both for training schemes and in the field.

Eschborn, June 1988

Dr. K. Erbel
Head of the Dept. for
Hydraulic Engineering
and Water Supply,
GTZ

# Preface

This laboratory manual is not intended to be just another introduction to the testing of water, waste water and soil. The aim is to fulfill the requirement for a comprehensive work, including descriptions of relatively simple chemical, microbiological and soil science techniques that can easily be carried out in technical laboratories. Methods involving complex apparatus have been deliberately omitted and only those soil science tests included which can be performed in a poorly to averagely equipped water laboratory included. Irrigation aspects and environmental protection are emphasized.

Experience with relevant projects in third world countries clearly shows that experimental details alone are unable to satisfy the requiremets of technical personnel working in the field. Management and laboratory personnel are often required to interpret their results on the basis of given reference values or technical specifications. The necessary assessment criteria are normally camouflaged within national regulations, technical and scientific regulations or standardization procedures. For this reason, part of the book deals exclusively with the interpretation of results.

In many laboratories, the quality of an analysis is not always up to satisfactory standard. Also, test results are not always employed correctly to influence technical practice in water, waste water and soil analysis. The creation of "grave-yards" of numbers is thus a common consequence. In order to illustrate possible ways of avoiding such problems, methods of laboratory quality control, data documentation and evaluation are covered in some depth.

Furthermore, test requirements for the evaluation and assessment of ground water, surface water, drinking water, waste water and soil are discussed so that the major test parameters can be included during the planning stages of measuring programs. The organization of sampling programs and an introduction to sampling techniques are treated in detail, as experience shows that poor quality in these areas cannot be compensated by good laboratory practice. Finally, questions of laboratory safety, essential requirements for all work with dangerous substances and equipment, are considered.

The following points are important for the user of the manual:
– Previous experience of practical laboratory work is assumed so that most procedures are described in a shortened form. On the other hand, the authors are of the opinion that the procedures described for the application of laboratory reagents and equipment are provided in sufficient detail that reference to less accessible texts can generally be dispensed with.
– Chemical formulae are only given where the name alone is not sufficient for unambiguous identification.
– As a rule chemical reagents of analytical quality should be employed.
– Deionized water should be used for the preparation of aqueous solutions unless otherwise specified.
– For the sake of clarity, the derivation and exact formulation of mathematical equations are generally omitted.

The authors wish to thank all colleagues and fellow employees who generously provided their professional advice in the preparation of this book.

Taunussstein and Eschborn  H. H. Rump
June 1988  H. Krist

# Contents

| | | |
|---|---|---|
| **1** | **Safety in the Laboratory** | 1 |
| 1.1 | Basic Rules for Laboratory Safety | 1 |
| 1.2 | Handling of Chemicals | 2 |
| 1.3 | Fire Hazards and Fire Prevention | 5 |
| 1.4 | Electricity | 7 |
| 1.5 | First Aid | 8 |
| 1.6 | Disposal of Dangerous Laboratory Waste | 9 |
| | | |
| **2** | **Quality Control** | 11 |
| 2.1 | General | 11 |
| 2.2 | Statistical Methods | 13 |
| 2.2.1 | Fundamental Principles | 13 |
| 2.2.2 | Background Quality Control | 21 |
| 2.2.3 | Routine Quality Control | 24 |
| 2.3 | Documentation of Analytical Results | 25 |
| | | |
| **3** | **Requirements for Analytical Methods** | 30 |
| 3.1 | Ground Water | 31 |
| 3.2 | Surface Water | 36 |
| 3.3 | Drinking Water | 38 |
| 3.4 | Waste Water | 40 |
| 3.5 | Soil | 42 |
| | | |
| **4** | **Organization of Sample Collection Program and Sampling Techniques** | 48 |
| 4.1 | General | 48 |
| 4.2 | Organization of Sampling Networks | 50 |
| 4.3 | Determination of Water Quality | 54 |
| 4.4 | Sampling Devices | 57 |
| 4.5 | Preservation, Transport and Storage of Samples | 60 |
| 4.6 | Sample Collection Procedure | 61 |
| 4.6.1 | Ground Water | 62 |
| 4.6.2 | Surface Water | 63 |
| 4.6.3 | Drinking Water | 63 |
| 4.6.4 | Waste Water | 63 |
| 4.6.5 | Soil | 64 |
| | | |
| **5** | **Field Measurements** | 66 |
| 5.1 | Check List | 66 |
| 5.2 | Parameters | 66 |
| 5.2.1 | Sensory Examination | 66 |
| 5.2.2 | Temperature | 68 |
| 5.2.3 | Settleable Matter | 69 |
| 5.2.4 | pH-Value | 69 |

| | | |
|---|---|---|
| 5.2.5 | Redox Potential | 71 |
| 5.2.6 | Electrical Conductivity | 72 |
| 5.2.7 | Oxygen | 73 |
| 5.2.8 | Chlorine | 76 |
| 5.2.9 | Alkalinity (Acidic Capacity) | 79 |
| 5.2.10 | Acidity (Base Capacity) | 80 |
| 5.2.11 | Calcium Carbonate Aggression | 82 |
| **6** | **Laboratory Measurements** | **85** |
| 6.1 | Sample Preparation | 85 |
| 6.1.1 | Water and Waste Water Samples | 85 |
| 6.1.2 | Soil Samples | 85 |
| 6.2 | Analytical Methods | 87 |
| 6.2.1 | Chemical Analysis | 87 |
| 6.2.1.1 | Ammonium | 87 |
| 6.2.1.2 | Biochemical Oxygen Demand | 89 |
| 6.2.1.3 | Boron | 92 |
| 6.2.1.4 | Calcium and Magnesium | 93 |
| 6.2.1.5 | Chemical Oxygen Demand | 95 |
| 6.2.1.6 | Chloride | 100 |
| 6.2.1.7 | Copper | 101 |
| 6.2.1.8 | Cyanides | 102 |
| 6.2.1.9 | Dissolved and Undissolved Substances | 106 |
| 6.2.1.10 | Iron | 109 |
| 6.2.1.11 | Kjeldahl Nitrogen | 111 |
| 6.2.1.12 | Manganese | 113 |
| 6.2.1.13 | Nitrate | 116 |
| 6.2.1.14 | Nitrite | 117 |
| 6.2.1.15 | Oils and Fats | 118 |
| 6.2.1.16 | Phenol Index | 120 |
| 6.2.1.17 | Phosphorus Compounds | 122 |
| 6.2.1.18 | Potassium | 125 |
| 6.2.1.19 | Silicic Acid | 127 |
| 6.2.1.20 | Sodium | 128 |
| 6.2.1.21 | Sulfate | 129 |
| 6.2.1.22 | Surfactants | 132 |
| 6.2.1.23 | Zinc | 134 |
| 6.2.2 | Microbiological Testing of Water | 135 |
| 6.2.2.1 | Sampling, Transport and Storage of Water Samples for Microbiological-hygienic Tests | 136 |
| 6.2.2.2 | Technical Requirements for the Hygienic Water Examination | 137 |
| 6.2.2.3 | Carrying out Microbiological Water Tests | 138 |
| 6.2.2.3.1 | Total Bacterial Count | 138 |
| 6.2.2.3.2 | Escherichia coli and Coliform Bacteria | 140 |
| 6.2.2.3.3 | Further Hygienically Important Microorganisms in Water | 144 |
| 6.2.2.4 | Preparation of Nutrient Solutions and Media | 146 |
| 6.2.3 | Specialized Soil Analyses | 150 |
| 6.2.3.1 | Mesh Size | 151 |

| | | |
|---|---|---|
| 6.2.3.2 | Hydrolytic Acidity (H-Value) | 152 |
| 6.2.3.3 | Exchangeable Basic Substances (s-Value) | 153 |
| 6.2.3.4 | Exchange Capacity | 154 |
| 6.2.3.5 | Carbonate Content | 154 |
| 6.2.3.6 | SAR-Value (Sodium Absorption Ratio) | 155 |
| **7** | **Interpretation of Test Results** | 158 |
| 7.1 | Ground Water | 158 |
| 7.2 | Surface Water | 158 |
| 7.3 | Drinking Water | 160 |
| 7.4 | Water for Use in Construction | 168 |
| 7.5 | Water for Irrigation | 169 |
| 7.6 | Waste Water | 172 |
| 7.7 | Soil | 174 |
| **Literature** | | 177 |
| **Appendix** | | 179 |
| | Statistical Tables | 179 |
| | BASIC-Programs | 183 |
| **Index** | | 187 |

# 1 Safety in the Laboratory

Sources of danger can never completely be avoided during work in chemical laboratories. However, the risk may be reduced by application of specialist knowledge in chemistry or physics, as well as an understanding of basic safety regulations for laboratory and industrial practice. The following chapter is designed to provide a simple introduction to such rules and should be thoroughly studied, especially by beginners and inexperienced personnel.

Typical sources of danger in the chemical laboratory are encountered when using:

- poisonous, flammable or explosive substances
- high temperatures and pressures
- electricity.

Specific accidents in the laboratory result from:

poisoning; fires and explosions through working with flammable gases, fumes or materials; contact with hot materials, heated liquids, acids and alkali; explosion of containers under pressure; the effects of electric current.

The fundamental rule of accident prevention must be kept in mind:

**NEVER BE INATTENTIVE,**

**NEGLIGENT OR CARELESS**

## 1.1 Basic Rules for Laboratory Safety

Work with laboratory chemicals and equipment becomes less dangerous when the following basic rules are applied:

only qualified personnel should be permitted to work with dangerous reagents and equipment.

- safety goggles must always be worn; in addition, protective gloves and other protective clothing should be worn where necessary.

- working areas must be well ventilated; an efficient fume-hood must be employed where dangerous gases or fumes are released.

- chemicals should not be permitted to come into contact with the eyes, mucous membranes or skin.

● contaminated eyes must be washed thoroughly with water or special wash bottles, the subject being in a reclining position.

● splashes of liquids and dangerous substances on the skin should first be removed with a dry cloth or tissue and the contaminated area held for some time under runnig water. Washing is then completed with warm water and soap.

● clothing contaminated with corrosive substances must be removed at once.

## 1.2 Handling of Chemicals

Many chemicals which can be toxic, detrimental to health, corrosive, irritating, flammable or highly flammable are a potential danger to the persons handling them.

Usage: Dangerous chemicals must not be stored in breakable containers exceeding 5 liters capacity. The only exceptions are special protection systems such as trapping containers in combination with absorptive and fire extinguishing materials

Special dangers associated with the analytical procedures described in this handbook are referred to by means of symbols (see Fig. 1).

Transport: Breakable containers should never be carried by the neck only but also supported underneath. For transport of containers over longer distances, safe handling must be ensured e.g. in buckets, baskets or wooden cases. The risk of slipping on condensed water when removing containers from cold rooms should be pointed out.

Storage: In general, laboratory chemicals should be stored under cool and dry conditions. Larger quantities of dangerous materials should never be kept in the laboratory itself but in a chemical store, constructed according to official regulations. Explosive and flammable substances must always be stored separately. Extra cold storage is recommended for the following:

● flammable gases in pressurized containers,

● liquids whose boiling points can easily be reached under unsuitable storage conditions e.g. acetone, diethyl ether, pentane, hexane, petroleum ether, carbon disulfide and dichloromethane.

## 1.2 Handling of Chemicals

highly toxic

highly flammable

potentially explosive

corrosive

**Fig. 1:** Hazard signs in chemical industry

If possible, the following should be stored in a lockable room or cupboard:

● substances with special danger symbols present on their containers e.g.:
cyanides
mercury and its compounds
arsenic and its compounds
alkali metals
thallium compounds
uranium compounds
phosphorus
pesticides

The following must be stored under well ventilated conditions:
- chemicals which release corrosive fumes e.g.:

hydrofluoric acid,

bromine,

nitric acid,

hydrochloric acid,

ammonia solutions.

Certain substances must be isolated from others (Table 1):

**Table 1:** Chemical substances to be isolated from others

| Substances | To be kept separate from |
|---|---|
| Active charcoal | oxidizing agents, calcium hypochlorite |
| Alkali metals | water, carbon dioxide, chlorinated hydrocarbons, halogens |
| Ammonia gas | mercury, halogens |
| Ammonium nitrate | acids, powdered metals, sulfur, flammable liquids, finely powdered organic materials |
| chlorates | ammonia salts, acids, powdered metals, sulfur, finely powdered organic materials |
| Chromium(VI)-oxide | flammable liquids |
| Copper | acetylene, hydrogen peroxide |
| Cyanides | acids |
| Flammable liquids | oxidizing agents such as ammonium nitrate, chromium(IV)-oxide, nitric acid, sodium peroxide, halogens |
| Hydrofluoric acid | ammonia |
| Hydrogen peroxide | metals and metal salts, organic compounds |
| Mercury | acetylene, ammonia |
| Nitric acid (conc.) | acetic acid, chromium(VI)-oxide, hydrogen sulfide, flammable liquids and gases |
| Oxalic acid | silver, mercury |
| Perchloric acid | alcohols, paper, wood, acetic anhydride, bismuth and its alloys |
| Phosphorus | sulfur, chlorates |
| Potassium permanganate | glycerol, ethylene glycol, sulphuric acid |
| Silver | acetylene, oxalic acid, tartaric acid, ammonium compounds |
| Sodium peroxide | flammable liquids |
| Sulfuric acid (conc.) | potassium chlorate, potassium perchlorate, potassium permanganate |

## 1.3 Fire Hazards and Fire Prevention

The following conditions must be satisfied for a fire to start:

- flammable material (solids, gases, fumes)
- oxygen, air (approx. 21% $O_2$)
- source of ignition (flame, spark, heating element or plate)

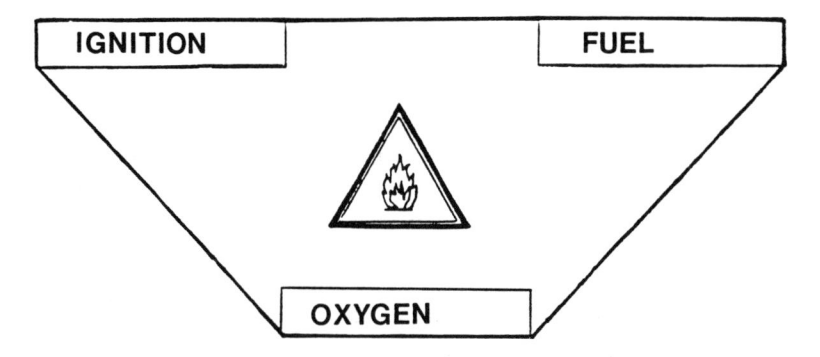

**Fig. 2:** Scheme of fire hazards

Flammable liquids can only combust at temperatures above the flash point and the fumes thereby produced are heated to above the ignition temperature. These both values are specific for a substance. The flash point of a flammable liquid is the lowest temperature at which a vapor/air mixture is formed over the liquid under defined conditions and can be ignited by external sources. The ignition temperature is the lowest temperature under well-defined experimental conditions and at normal pressure at which the flammable material spontaneously ignites.

The published values for flash points and ignition temperatures can be easily obtained from suitable reference works (see literature list). Substances with a flash point below 21 °C and which are therefore extremely flammable include:

- immiscible with water:

petrol, benzene, diethyl ether, carbon disulfide, ethyl acetate, toluene;

- miscible with water:

methanol, ethanol, propanol, isopropanol, pyridine, acetone, tetrahydrofuran.

Liquids with flash points between 21 and 55 °C:
butanol, butyl acetate, chlorobenzene, amyl alcohol, acetic anhydride, xylenes.

Liquids with flash points between 55 and 100 °C:
dichlorobenzene, cresols, heating oil, nitrobenzene, phenol, paraffin oil.

For storage of these liquids refer to Section 1.2

Mixtures of flammable materials with air can not be ignited in all proportions. Ignitable mixtures are characterized by their upper and lower explosive limits given as the volume fraction (%) or in $g/m^3$ (see literature list).

Rules for fire prevention:

● special care during heating of more than 50 mL inflammable liquid.

● the spreading of flammable vapors must be prevented. Vapors are often heavier than air and can spread over several metres in a short time. Sources of ignition (bunsen burners, heating plates and mantles) even when concealed or at some distance from the vapors should be removed.

● electrostatic charges can start fires through sparking. Charging can occur during filling of glass or plastic containers with non-conducting liquids e.g. acetone, ether, carbon disulfide, toluene. Therefore, fluids should be poured slowly avoiding free-fall and if possible using a suitable funnel reaching the container bottom. Conducting containers should be earthed. This also applies to containers and equipment which are either conducting or nonconducting.

● no easily ignitable liquid should be stored in a refrigerator since an ignition of gases may be caused by sparking (light, thermostat).

● bumping should be avoided during distillation by using anti bumping pellets or capillaries in the case of vacuum distillation.

Protective equipment and fire extinguishers:

● escape routes must be provided, clearly marked and kept free from hindrances. At least one second exit must be available,
  ● fire alarms must be installed and the emergency number clearly marked near the telephone,
  ● at least one safety shower per laboratory should be provided,
  ● fire-proof blankets must be readily available,
  ● availability of sand buckets,
  ● portable carbon dioxide extinguishing equipment or hand fire extinguishers of various types ($CO_2$, powder) are to be within easy reach,

- regular fire practice.

All fire protection equipment and materials including fire alarms are to be marked in red.

Fire fighting procedure:

- remove injured persons from the fire area,
- persons with burning clothing should be rapidly wrapped in an extinguishing blanket on the floor, or sprayed with carbon dioxide (not in the face!), or pulled under the laboratory shower,
- fire alarm,
- if possible, flammable material or gas cylinders should be removed,
- isolate gas supply at mains,
- in case of fire involving electrical equipment switch off electricity supply, if possible, before beginning fire fighting procedures,
- try to extinguish the fire or, in a hopeless situation, evacuate the building.

Fire fighting tips for the laboratory:

In most laboratory fires, carbon dioxide or Halon$^R$ extinguishers are sufficient. They leave no residue and therefore cause no contamination of rooms or damage to sensitive apparatus, are chemically neutral and can be used with electrical equipment. After successfully extinguishing a fire with carbon dioxide, the room must be immediately ventilated in order to avoid the danger of suffocation. Fires involving alkali metals or lithium aluminum hydride should under no circumstances be fought with water or Halon$^R$ extinguishers. In this case, cement powder is recommended. Carbon dioxide or powder extinguishers should be employed in flammable liquid fires.

## 1.4 Electricity

The widespread use of electrical equipment in the chemical laboratory should cause much attention to be paid to related sources of danger. Severe burns can result when the human body come into contact with current-carrying equipment. In addition, the heart-beat can be shocked out of rhythm. The current which is caused to flow through the body is of decisive importance in determining the degree of damage caused.

The greater the contact potential and the lower the resistance, the greater is the resulting current. Thus, with wet hands and conducting floor even potentials as small as 50 V can be dangerous, since large currents are attained. The duration of current flow is also of importance in assessing the damage caused. The danger caused by contact with electrical equipment is illustrated by the following example:

The average resistance of the human body is $1\,300\,\Omega$. In the case of a good contact e.g. on hands and feet, a current of approximately 170 mA flows through the body at a potential of 220 V. This current can cause death within several seconds. These facts apply to A. C. supplies. In the case of direct current, special care should be exercised since relatively small potentials can be fatal due to the resulting electrolytic processes in the body.

The following rules must be observed when working with electricity:

● all equipment with more than 50 V potential is of the heavy current type. Installation and repairs should only be carried out by properly qualified specialists according to national regulations.

● sockets, plugs, cables and equipment should be tested regarding isolation before use. Defective components as well as wet equipment represent potential dangers.

● the electricity supply in each room should be easily cut off via a master switch placed in an accessible position.

● all apparatus should be equipped with earthed contact fuses. These should be tested from time to time.

### 1.5 First Aid

Here, only the more important points can be made. Under no circumstances should a practical course in first aid procedures be dispensed with. First aid given by amateurs is no substitute for medical treatment but is only an emergency step before a physician can take over.

Equipment of a laboratory for first aid:

● wall chart of first aid instructions,
● addresses and telephone numbers of emergency physicians and hospitals,
● first aid kit,
● eye wash bottles.

Procedure:

Corrosive burns: The affected parts of skin are washed throughly with water.

Bromine burns: Wash with paraffin oil or ethanol.

Hydrofluoric acid burns: Wash with 2% ammonia solution or dilute sodium bicarbonate solution.

Iodine burns: Wash with 1% sodium thiosulfate solution.

Eye burns: Hold the affected eye wide open with both hands and wash for about 10 minutes with water. The eye should be rolled in all directions during washing.

Cuts: Do not touch or wash the wound but cover with sterile material (plaster, bandage).

Burns: Hold affected parts for up to 15 minutes in cold water. Burns are to be kept sterile and covered with a special burn dressing (not for facial areas). Burn wounds caused by phosphorus should be bathed with sodium bicarbonate solution.

Poisoning: If conscious, vomiting should be induced. Activated charcoal tablets may also be given. Any remaining poisonous and/or vomited material must be kept for possible examination. Lie the victim on his/her side.

Where unconscious:

- lie the subject on his/her side,
- bend head backwards, face towards the floor,
- monitor pulse and breathing,
- if breathing is interrupted: first of all blow air strongly 20 x into the mouth or nose, then wait for approx. 30 seconds and continue with normal artificial respiration procedures. According to experience and circumstances begin with heart massage if necessary. These steps should be maintained until the physician arrives.

## 1.6 Disposal of Dangerous Laboratory Waste

Dangerous waste must be collected in such a way that any risk for personnel is entirely eliminated. The following rules are to be observed:

- the disposal can only be carried out by experienced personnel or specialist organisations.

- under no circumstances are cyanides to be washed into the drainage system. They must be caused to react, e.g. with iron salts.

- waste materials which can produce poisonous or easily combustible gases and fumes, or which react with water (sodium, potassium, carbides, phosphides) are to be collected in fire-resistant containers for subsequent disposal.

- flammable liquids should be collected and under no circumstances disposed of in the drainage system. Chlorinated hydrocarbons should be collected separately.

- solutions containing heavy metals should never be directly disposed of in the drainage system but treated in the laboratory by "filtering" through granulated magnesium oxide/marble (1:1). This mixture can be placed in a broad shallow column (prepared e.g. by cutting a polythene bottle) and fixed in a convenient position over the sink.

- acids and alkalis should either be neutralized centrally in a special plant or in small amounts in a suitable container. The process may be monitored by pH-paper or by a pH-meter.

# 2 Quality Control

## 2.1 General

The quality of laboratory test results is of paramount importance for all conclusions drawn from them. Inaccurate or false results can lead to serious errors in judgement.

Quality control concerns those activities which have to do with the measurement and minimization of possible errors.

Although much effort has been put into standardizing analytical measuring procedures (ISO, ASTM, DIN), only few generally recognized methods exist regarding quality control (mainly statistical techniques).

Figure 3 below illustrates the problems encountered in judging and controlling errors of measurement. Two types of errors can be identified:

- systematic errors
- random errors

a) large proportion random errors
   no systematic errors

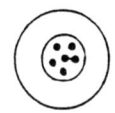

b) small proportion random errors
   no systematic errors

c) small proportion random errors
   large proportion systematic errors

d) large proportion random errors
   large proportion systematic errors

**Fig. 3:** Type of errors in laboratory measurements

Reliable and comparable analytical results can only be obtained through the introduction of a comprehensive quality control system accommo-dating all steps and procedures involved in the chemical laboratory.

Quality control should therefore begin during sample collection and be carried through to the final preparation and documentation of results. The aims are:

- to obtain the highest accuracy and precision of the analytical results.

- to ensure that the reliability of analysis i.e. accuracy of results is maintained in the future.

**Only those results having a known accuracy and precision can be compared.**

These general statements should find application in the organisation of every laboratory, large or small. Between 10 % and 20 % of all routine tests should be subjected to quality control procedures. This requirement may seem to be excessive but it is certainly better to produce a smaller number of accurate analytical results than a large number of results containing undefined errors.

The complete avoidance of analytical errors in the laboratory is clearly impossible. In this chapter a range of simple techniques will be described which allows error control to be improved. It should always be clear that the most important factor is the laboratory technician himself. Without a self-critical approach, all quality control techniques are worthless.

Quality control of routine measurements via application of statistical techniques consists of two parts:

- background quality control,

- routine quality control.

The background quality control includes choice of the analytical technique and determination of its advantages and suitability. The routine quality control includes general internal and external laboratory quality control.

The general rules of "Good Laboratory Practice (GLP)" can be referred to in carrying out quality control. Even where all such rules are not implemented, it is recommended that suitably qualified staff members be assigned the task of introducing and running a quality control system. Every laboratory should have a quality control officer (full or part-time, according to laboratory size) who must not be head of the laboratory in question. All such examination results are discussed with the relevant personnel in order to resolve any problems found. The examination reports are constantly recorded and filed.

## 2.2 Statistical Methods

### 2.2.1 Fundamental Principles

Where repeated analyses are carried out on the same homogeneous sample, identical analytical results are not obtained each time but various results whose distribution can be represented by a histogram. A large number of such repeated measurements allows construction of a distribution curve which can often be a normal Gaussian curve (Figure 4).

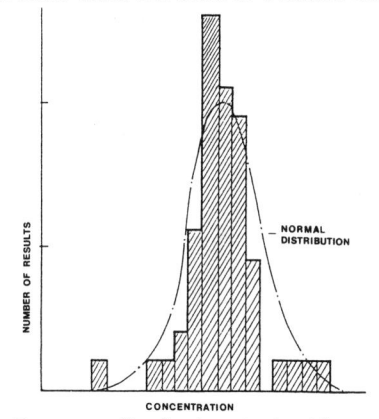

**Fig. 4:** Frequency distribution obtained from repeated analyses of the same sample

This curve represents the relationship between the numerical value of an analytical result and its probability of occurrence. The distribution of the probability about its maximum is symmetrical. The main parameters describing such a distribution are the mean, x, and the standard deviation, s. Both values are normally calculated during evaluation of measurements. The mean is defined as

$$\bar{x} = \frac{x_1 + x_2 + \ldots x_N}{N}$$

where $x_1, x_2 \ldots x_N$ are the values to be considered,

The standard deviation s (whose formula will not be derived here) is:

$$s = \pm \sqrt{\frac{\Sigma(f_1^2 + f_2^2 + \ldots f_N^2)}{N-1}}$$

where    f  deviation of a single measurement from the mean
         N  number of single measurements

Example:  Calculation of $\bar{x}$ and s:

Different volumes of titrant were consumed during repeated titrations of the same sample. Mean, $\bar{x}$ and standard deviation, s are calculated as follows:

**Table 2:** Calculation of mean and standard deviation

| mL x consumed | $\bar{x}$ mL | $x - \bar{x} = f$ mL | $f^2$ $mL^2$ |
|---|---|---|---|
| 21.33 | 21.37 | -0.04 | 0.0016 |
| 21.30 | 21.37 | -0.07 | 0.0049 |
| 21.34 | 21.37 | -0.03 | 0.0009 |
| 21.45 | 21.37 | +0.08 | 0.0064 |
| 21.42 | 21.37 | +0.05 | 0.0025 |
| $\sum x = 106.84$ | | | $\sum f^2 = 0.0163$ |

$$\bar{x} = \frac{106.84}{5} = 21.37 \text{ ml}$$

$$s = \pm \sqrt{\frac{0.0163}{4}} = \pm 0.06 \text{ ml}$$

The standard deviation, s, is a statistical parameter of the range into which measurements of a series fall. If a normal distribution exists

- approximately 68 % of values lie within one standard deviation of the mean, i.e. $\bar{x} \pm s$,
- approximately 95 % of values lie within two standard deviations, i.e. $\bar{x} \pm 2s$,
- approximately 99.7 % of values lie within three standard deviations, i.e. $\bar{x} \pm 3s$.

These values are obtained by integration of the corresponding areas under the curve shown in Figure 5.

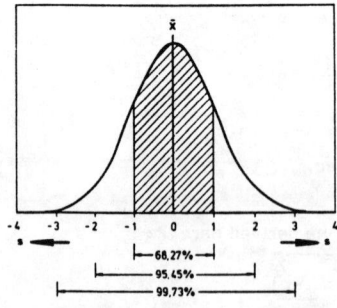

**Fig. 5:** Areas corresponding to 1-3 standard deviations (normal distribution assumed)

## 2.2 Statistical Methods

For our example given above, the following is valid:

- approx. 68 % of values lie in the region of 21.37 mL ± 0.06 mL i.e. from 21.31 mL to 21.43 mL,

- approx. 95 % of values lie in the region between 21.37 mL ± 2  0.06 mL i.e. from 21.25 mL to 21.49 mL.

Sometimes, it is advantageous to give the standard deviation, s, relative to the mean $\bar{x}$. This relative standard deviation is known as the relative error or the coefficient of variation, V, and is calculated according to:

$$V = \frac{s}{\bar{x}} \cdot 100 \ \%$$

For the above example:

$$V = \frac{\pm 0.06 \text{ mL}}{21.37 \text{ mL}} \cdot 100 \ \% = 0.3 \ \%$$

However, analytical results are not always distributed normally. Other distributions are encountered e.g. during evaluation of time-dependent water source data (e.g. river water, analyses including flood water samples) or in measurement of the same sample by various laboratories (e.g. inter-laboratory trials) (Figure 6). An indication of non-normally distributed data is when V > 100 %.

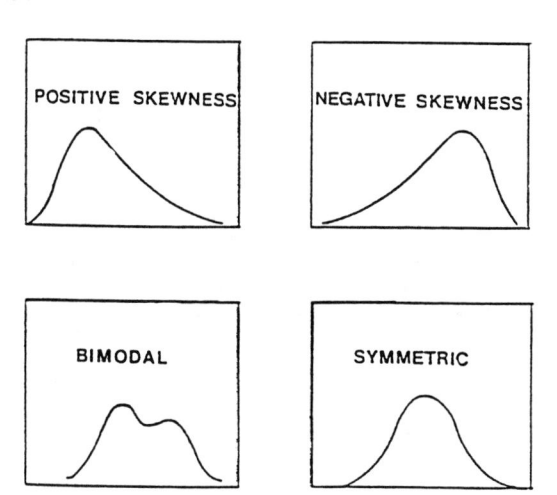

**Fig. 6:** Characteristic types of distribution curves

Characteristic examples of distribution curves (right skewness = positive skewness, left skewness = negative skewness).

The curve shape gives an indication as to the presence of non - compatible statistical information e.g. various water types, or to systematic errors within the measurements. In such cases, the data must be separately examined statistically, since mean and standard deviation are less informative parameters in such cases. Such an example is shown in Figure 7.

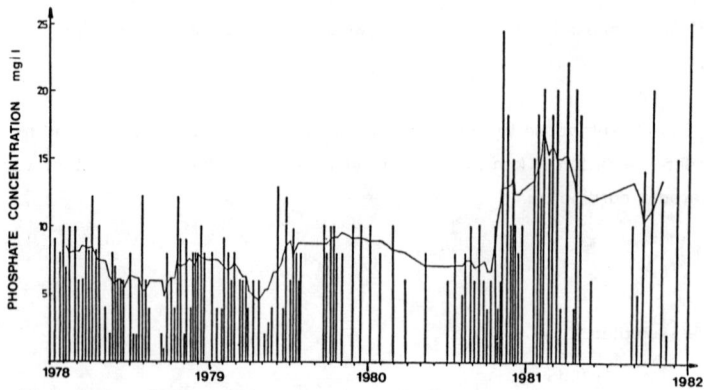

**Fig. 7:** Time Series of the phosphate concentration in a river

The whole series of measurements consists of two basic groups since after about 1981 significantly higher values were measured. The continuous curve represents the mean values, each calculated from 5 single measurements. A single statistical treatment of all values would give a false description of the data.

Trends also lead to difficulties in interpreting data, as shown in Figure 8.

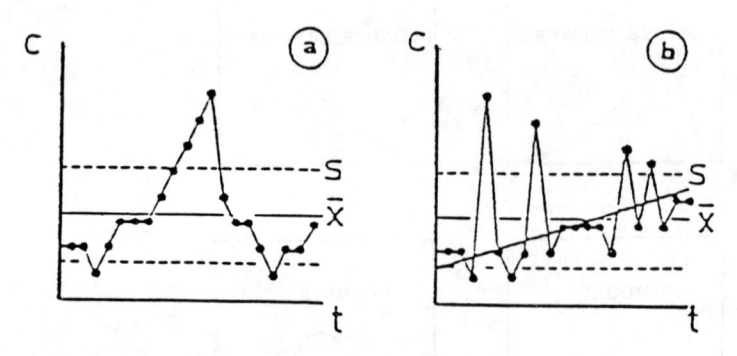

**Fig. 8:** Time Series of concentrations (a) cyclic, (b) trend

Both situations a and b have the same mean and standard deviation although they consist of measurements exhibiting a) periodical variation and b) a trend. It is therefore clear that further statistical techniques and aids such as graphs cannot be dispensed with.

Before a further statistical treatment of laboratory data is attempted, it is important to identify and eliminate stray values since many statistical techniques are based upon a given distribution e.g. normal distribution. It is therefore possible that a threshold appears to be exceeded due to the occurrence of a single non-representative value.

Statistical tests have been developed to provide objective clarification of such situations. Set equations are employed to calculate test parameters which may then be compared with values published in tables. Important test methods include:

### DIXON-test for stray values

The measured values of a sample are ordered according to size and the following equations applied:

$$\text{for } N \leq 7 \qquad Q = \frac{x_1 - x_2}{x_1 - x_N}$$

$$\text{for } N \geq 7 \qquad Q = \frac{x_1 - x_2}{x_1 - x_{N-1}}$$

Here, $x_1$ is the suspected deviant, $x_2$ the neighbouring value in the series and $x_N$ the value at the other end of the series. The test parameter, $Q$, is then compared with the corresponding table value at a 95% significance level (see Appendix). If $Q$ exceeds the table value, the tested value $x_1$ is a stray value which must be eliminated. For the following tests a normal distribution is obligatory. Non-normally distributed data sets may be transformed into normally distributed ones (e.g. logarithmic transformation).

### F-Test

This test is designed to compare two standard deviations. In this way, the superiority of two methods, laboratories or staff members can be evaluated.

$$\text{Test equation:} \qquad F = \frac{s_1^2}{s_2^2}$$

where the larger value for the variation $s^2$ must be the numerator. The parameter F is usually tested at a 95% level of significance. When F is larger than the table value (see Appendix), the standard deviations are different. Here

$$f_1 = N_1 - 1 \quad \text{and} \quad f_2 = N_2 - 1$$

### t-Test

The t-test is employed to compare two means with the same or different variances. Two different tests can be carried out:

### a) one-sample t-test

This test compares a mean with a real value and therefore leads to detection of systematic errors.

$$t = \frac{|\bar{x} - \tilde{x}|}{s} \cdot \sqrt{N}$$

The test parameter, t, is usually compared with corresponding table values at significance levels of 95% (see Appendix). Where t is greater than the table value, the mean has a systematic error. Here $f = N - 1$.

### b) two-sample t-test

This test examines differences between two means $\bar{x}_1$ and $\bar{x}_2$ from two different series of analyses of the same sample.

$$t = \frac{|\bar{x}_1 - \bar{x}_2|}{s_d} \cdot \sqrt{\frac{N_1 N_2}{N_1 + N_2}}$$

$$\text{with } s_d = \sqrt{\frac{(N_1 - 1) s_1^2 + (N_2 - 1) s_2^2}{N_1 + N_2 - 2}}$$

A calculated value exceeding that in the table (see Appendix) reveals a difference in mean values at the 95 % significance level and thus, both means originate from different populations. Here, $f = N_1 + N_2 - 2$.

When examining the relationship between two series of measurements, the fundamental difference between functional and stochastic (statistical) relationships must be considered. A functional relationship can be reversed and defined analytically for each pair of values. A statistical relationship is also reversible but cannot be so clearly defined. Figure 9 is designed to clarify this:

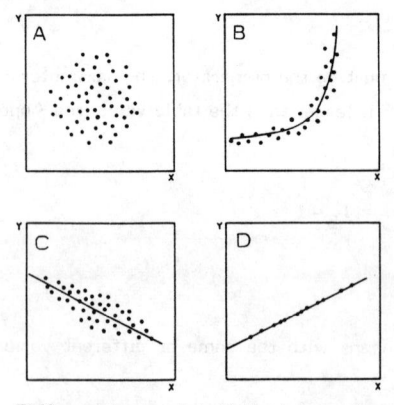

**Fig. 9:** Different types of statistical dependance of two variables

## 2.2 Statistical Methods

In A, no relationship is immediately obvious; in B, a higher order relationship is clear; in C, a negative statistical relationship is given and D shows a positive linear functional relationship.

The degree of a linear statistical relationship can be investigated by correlation analysis. The degree of association is given by the correlation coefficient and type of relationship by the slope of the fitted curve. The equation representing this plot is determined by regression analysis employing the method of least squares. Both techniques are commonly used to prepare and evaluate analytical calibration graphs. The corresponding equations are:

Correlation:

$$r = \frac{\sum(x_i - \bar{x}) \cdot (y_i - \bar{y})}{\sqrt{\sum(x_i - \bar{x})^2 \cdot \sum(y_i - \bar{y})^2}}$$

r always lies between +1 and -1. When r = 0, the measurements are independent of each other. When r = $\pm$ 1, the correlation is a strictly linear functional one.

Regression:

y depends on x :

$$y = a + b x \qquad \text{where:}$$

$$b = \frac{\sum(x \cdot y) - \sum x \cdot \sum y / n}{\sum x^2 - (\sum x)^2 / n}$$

$$a = \bar{y} - b \bar{x}$$

Example:

A calibration curve is to be calculated (in additional to the usual graphic representation) from the concentrations of known standard solutions and their respective photometric extinction values giving the regression parameters and correlation coefficient (Tab. 3). It is important to know with N<5 a calculation of the correlation is futile.

**Table 3:** Calculation of the regression grades and correlation coefficients for a calibration curve

| Concentration x | Extinction y | $x^2$ | $x \cdot y$ | $x - \bar{x}$ | $(x - \bar{x})^2$ | $y - \bar{y}$ | $(y - \bar{y})^2$ | $(x - \bar{x}) \cdot (y - \bar{y})$ |
|---|---|---|---|---|---|---|---|---|
| 0  | 0.030 | 0   | 0      | -15 | 225 | -0.301 | 0.093 | 4.515 |
| 5  | 0.132 | 25  | 0.660  | -10 | 100 | -0.199 | 0.040 | 1.990 |
| 10 | 0.236 | 100 | 2.360  | -5  | 25  | -0.095 | 0.009 | 0.475 |
| 15 | 0.335 | 225 | 5.025  | 0   | 0   | 0.004  | 0     | 0     |
| 20 | 0.419 | 400 | 8.380  | 5   | 25  | 0.088  | 0.008 | 0.440 |
| 25 | 0.542 | 625 | 13.550 | 10  | 100 | 0.211  | 0.045 | 2.110 |
| 30 | 0.623 | 900 | 18.690 | 15  | 225 | 0.292  | 0.085 | 4.380 |

$\sum = 105$  $\sum = 2.317$  $\sum = 2275$  $\sum = 48.665$  $\sum = 700$  $\sum = 0.280$  $\sum = 13.910$

$\bar{x} = 15$  $\bar{y} = 0.331$

$$b = \frac{48.665 - 105 \cdot 2.317/7}{2275 - 105^2/7} = \frac{13.91}{700} = 0.01987$$

$$a = 0.331 - 0.01987 \cdot 15 = 0.033$$

The regression equation is then:

$$y = 0.033 + 0.01987 \cdot x$$

correlation coefficient

$$r = \frac{13.910}{\sqrt{700 \cdot 0.280}} = \frac{13.910}{14} = 0.9936$$

The results reveal that the linear regression of such a calibration curve is very strongly functional.

## 2.2.2 Background Quality Control

Background quality control covers all activities which contribute to the determination, testing and improvement in the quality of analytical techniques, calibration standards, reagents and specimens. Other factors which play an important role include:

- a description of the analytical method,
- adjusting and calibration of equipment,
- calibration of the analytical method.

The choice of a suitable technique with its related sources of errors is of very great importance for preparatory work in quality control. Many methods in analytical chemistry require calibration procedures i.e. a calibration function must be calculated to enable concentrations to be derived from the measurements made. In the simplest cases calibration standards, concentrations and their measurement values can be plotted graphically. The calibration curve can then be fitted by eye. However, such a procedure often leads to errors and thus the calculation of linear regression is strongly recommended. The calibration function

$$y = a + b \cdot x$$

where a is the calculated zero value and b the slope (representing the sensitivity of the measurement technique) depicts the linear plot with the smallest possible deviation from the experimental values (computer program No. 3 in the Appendix). The linear regression model is based on three assumptions:

- linearity over a wide range,
- constancy of variation over the whole range,
- normal distribution of data.

The actual calibration graph lies within a range of reliability (Figure 10) (confidence interval VB).

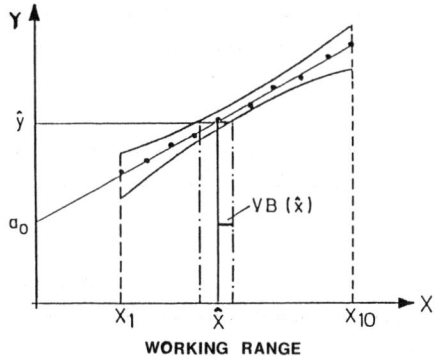

**Fig. 10:** Reliability of a calibration curve

This limited reliability range depends, among other things, on the deviation of the calibration points from the curve (residual standard deviation $s_y$) and the slope b. A parameter which shows the quality of a calibration curve is the methodological standard deviation $s_{xo}$. This is calculated by

$$s_{xo} = \frac{s_y}{b}$$

The methodological standard deviation may be employed to compare analytical procedures over the same working range with the same numbers and positions of calibration points. An example is shown in Figure 11.

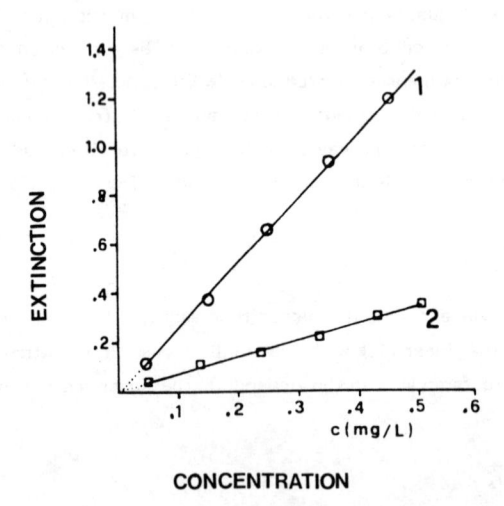

**Fig. 11:** Photometric determination of nitrite:
1 = sulfanilamide + N-(1-naphthyl)-ethylendiamine
2 = 4-aminosalicylic acid + 1-naphthol

It is clear that the residual standard deviations $s_y$ are similar but the sensitivities differ. A numerical comparison of both $s_y$ values with the F-test reveals that the two procedures differ considerably.

Methodological standard deviations are by no means readily available for all analytical techniques so that an internal laboratory examination is necessary. The (following) scheme illustrates the comparison strategy (Figure 12).

It must be remembered that every standard solution must be investigated using the same analytical procedure (including digestion where necessary) as the actual samples.

## 2.2 Statistical Methods

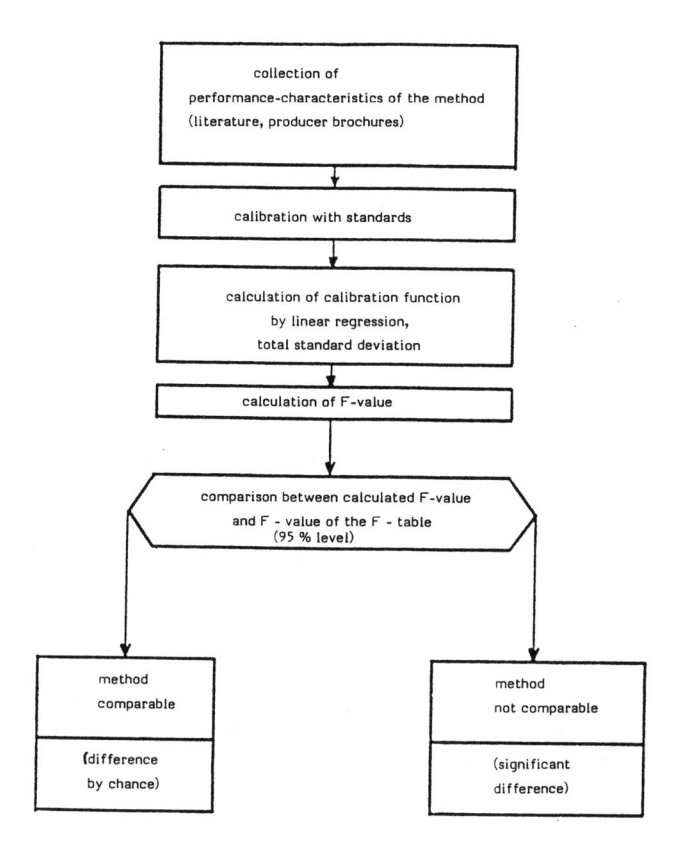

```
┌─────────────────────────────────┐
│         collection of           │
│ performance-characteristics of  │
│          the method             │
│   (literature, producer         │
│         brochures)              │
└────────────┬────────────────────┘
             ▼
┌─────────────────────────────────┐
│    calibration with standards   │
└────────────┬────────────────────┘
             ▼
┌─────────────────────────────────┐
│ calculation of calibration      │
│ function by linear regression,  │
│    total standard deviation     │
└────────────┬────────────────────┘
             ▼
┌─────────────────────────────────┐
│     calculation of F-value      │
└────────────┬────────────────────┘
             ▼
  ╱─────────────────────────────╲
 ╱ comparison between calculated ╲
│      F-value and F-value of     │
 ╲      the F-table (95 % level) ╱
  ╲─────────────────────────────╱
       │                    │
       ▼                    ▼
┌──────────────┐     ┌──────────────┐
│   method     │     │    method    │
│  comparable  │     │ not comparable│
├──────────────┤     ├──────────────┤
│ (difference  │     │ (significant │
│  by chance)  │     │  difference) │
└──────────────┘     └──────────────┘
```

**Fig. 12:** Strategy for the comparison of analytical methods

The methodological standard deviation also serves as a laboratory self-control:

- to compare measured with ideal values (e.g. during initial experiments),
- to train new staff,
- to optimize the various stages of a procedure,
- to test new analytical equipment.

As a minimum requirement for background quality control, it is recommended that the laboratory determinations (either single or double) are repeated on several consecutive days. This should include determination of:

- blank value,
- standard solutions at the high and low concentration limits of the defined working range,
- real samples,
- real samples after spiking.

Observations should be recorded over a phenol of 10 to 20 days. Afterwards, the stability of the system is tested by calculating the values:
- $s_w$ = standard deviation within a given batch,
- $s_b$ = standard deviation between batches.

The values of $s_w$ and $s_b$ are then compared using the F-test at a pre-determined significance level (e.g. 95 %). Only when this test reveals a significant difference between $s_w$ and $s_b$ must a separate calibration be carried out for each analytical series.

### 2.2.3 Routine Quality Control

After the preliminary tests have indicated that the background errors in the analytical procedure are small, the routine controls are implemented. A simple and established method is the use of control charts. Their application is based upon the assumption that the experimental data have normally distributed errors. For each control sample (blank, standard or real), the mean, $\bar{x}$, and standard deviation, s, are calculated. The control chart is then constructed with the mean as its center line. Warning and control areas are then added at $\pm 2$ s and $\pm 3$ s, respectively (see Figure 13a).

The control chart is normally followed over a period of one month. The analytical procedure judged to be under control when the measured values all fall within $\pm 2$ s. The following errors are possible when the procedure is out of control (Figure 13b).

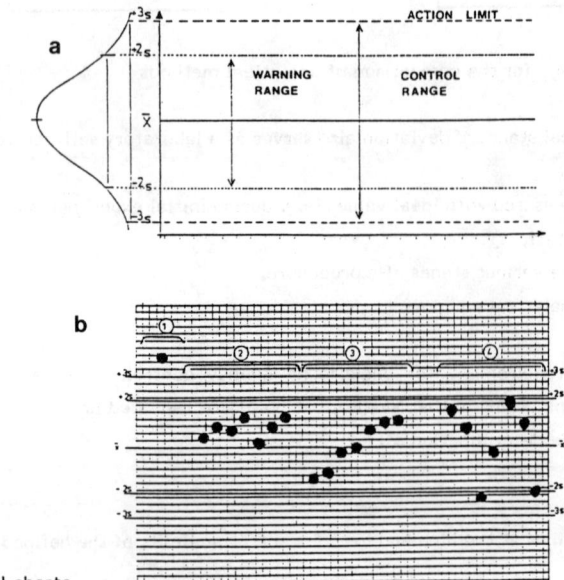

**Fig. 13:** Control charts
a) working ranges
b) detectable errors (see text)

The different kinds of errors can be characterized as follows:

1.  the values outside the warning range reveal gross analytical errors e.g. errors in the preparation of standards, presence of impurities or incorrect calibration.

2.  at least seven sequential results all lying above or below the mean e.g. use of new standards of a different quality.

3.  at least seven values showing an increasing or decreasing tendency e.g. degradation of the standard with age, or evaporation of solvent.

4.  sudden increased error e.g. technical error or insufficiently experienced personnel.

Such control charts can also be employed as a permanent record of the controls (for the limited control of spiked samples).

A further possibility is to use range control charts in which the differences in daily double determinations are plotted. This is recommended for controlling the precision of measurements carried out on real samples when it is expected that significant matrix effects are likely to have an influence on the outcome of the analysis. The control range of such a chart can, for example, be chosen to be 3.5 times the value of the mean variation from a previous period.

All of these internal quality control procedures describe the "reliability" (= precision and to a limited extent the accuracy) of routine analytical results and should be carried out for all parameters. Often such an effort directed towards quality control only becomes appreciated when the laboratory is involved in legal proceedings and has to provide evidence of its capability and the reliability of its analytical results.

Analytical results which have been subjected to internal quality control should be camparable to results from other laboratories. This degree of similarity can be tested by inter-laboratory trials. Such an external quality control should ensure that particularly the systematic errors are small and that the analytical results are acceptable. An inter-laboratory trial consists in the simplest case of analyses of standard samples. However inclusion of representative real samples is to be recommended. For example, when a minimum tolerable relative error of $\pm$ 10 % has been agreed upon, each laboratory must produce results lying within this range.

Details of performing inter-laboratory trials are to be found in technical literature (e.g. Cheeseman and Wilson, see Literature list).

## 2.3 Documentation of Analytical Results

The presentation of analytical results plays an important role in quality control since errors can often be more easily detected according to the chosen from of display, or similarly even the plausibility of complete analysis series can be tested. The origin of any form of presentation is the original data list e.g. in the form of a suitable analytical record. Figure 14 shows a form successfully employed for recording measurements made during a standard chemical-technical analysis e.g. of drinking water or in an assessment of the corrosive properties of water. The determination limits represent the lowest concentration of a substance that can be quantitively determined using a given procedure. The value differs significantly from 0. The level of detection of a procedure is defined as the smallest value which can be distinguished from the background measurements carried out on a control sample.

A form of presentation which is often optically very effective is the time-series graph. In Figure 15, the nitrite concentrations of surface water over 12 months are displayed both linearly (left side) and logarithmically (right side). In this way, time-series of measurements may be quickly compared.

**Designation**

| | | | |
|---|---|---|---|
| Date sampled: | | Sample taken by: | |
| Place sampled: | | | |
| Appearance: | | | |
| Water temperature: | °C | | |
| Air temperature: | °C | | |
| Conductivity: | mS/cm | | |
| Redox potential: | mV | | |
| pH-value at °C: | | | |
| pH-value after $CaCO_3$ saturation at °C: | | | |
| pH difference: | | | |

| | Determination limit | | Measured value | |
|---|---|---|---|---|
| Dissolved oxygen ($O_2$): | 0.1 | mg/L | | mg/L |
| Oxygen saturation index: | | | | % |
| Free chlorine ($Cl_2$): | 0.02 | mg/L | | mg/L |
| Bound chlorine ($Cl_2$): | 0.02 | mg/L | | mg/L |
| Acidity (base capacity) to pH 8.2: | 0.05 | mmol/L | | |
| as dissolved carbon dioxide ($CO_2$): | 2 | mg/L | | mg/L |
| Alkalinity (acid capacity) to pH 4.3 (m-value): | 0.05 | mmol/L | | |
| as hydrogen carbonate ($HCO_3^-$): | 3 | mg/L | | mg/L |
| Alkalinity (acid capacity) to pH 8.2 (p-value): | 0.05 | mmol/L | | |
| as carbonate ($CO_3^{2-}$): | 2 | mg/L | | mg/L |
| Calcium ($Ca^{2+}$): | 3 | mg/L | | mg/L |
| Magnesium ($Mg^{2+}$): | 1 | mg/L | | mg/L |
| Total alkaline earths (hardness): | 0.11 | mmol/L | | |
| Iron, total ($Fe^{2+/3+}$): | 0.02 | mg/L | | mg/L |
| Manganese ($Mn^{2+}$): | 0.02 | mg/L | | mg/L |
| Ammonium ($NH_4^+$): | 0.02 | mg/L | | mg/L |
| Nitrite ($NO_2^-$): | 0.02 | mg/L | | mg/L |
| Nitrate ($NO_3^-$): | 1 | mg/L | | mg/L |
| Chloride ($Cl^-$): | 3 | mg/L | | mg/L |
| Sulphate ($SO_4^{2-}$): | 2 | mg/L | | mg/L |
| Phosphate, total (as $PO_4^{3-}$): | 0.04 | mg/L | | mg/L |
| Oxidizability a) as potassium permanganate consumed | 1 | mg/L | | mg/L |
| b) as oxygen ($O_2$): | 0.3 | mg/L | | mg/L |

**Fig. 14:** Protocol sheet of a "chemical-technical standard analysis"

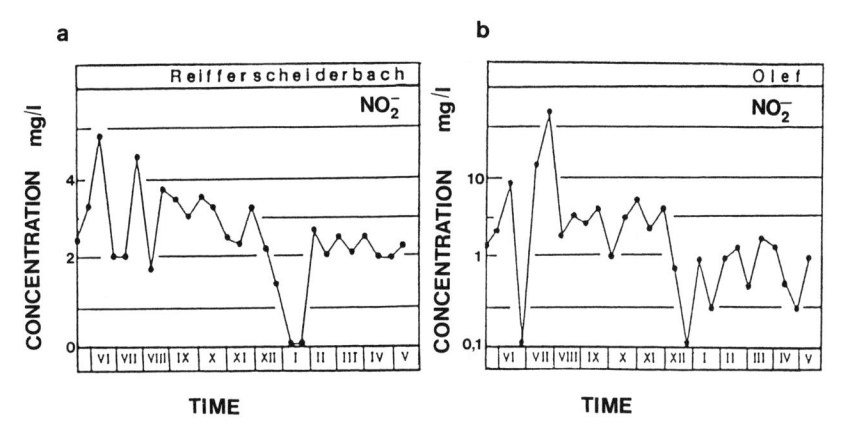

**Fig. 15:** Time-series of nitrite concentration in linear (a) and logarithmic (b) form

The form of representation shown in Figure 16 is recommended where series of measurements have been carried out at various points and a rapid overview of means, standard deviations and value ranges is required.

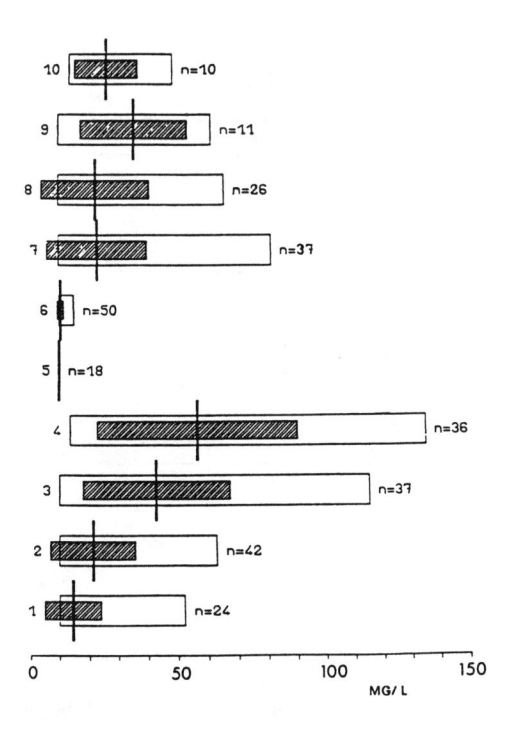

**Fig. 16:** Presentation of mean, standard deviation and range of ammonia concentrations in 10 waste water samples

The measurement points are indicated on the left and standard deviations and ranges represented by the horizontal bars. The number of measurements in each case is given beside each point in order to facilitate statistical tests of significance. A computer program (Program 1) for the calculation of means and standard deviations is listed in the Appendix.

In those cases where larger numbers of various parameters have been determined at one sampling point, possible inter-relationships may be documented in the form of a correlation matrix (in cases of non autocorrelated data). Such a matrix is shown in Figure 17 for various measurements on seepage water from a refuse dump. Close relationships are apparent between COD and BOD or between $NH_4^+$ and COD. Significance must be tested before such correlation coefficients can be evaluated. A table for testing the significance is listed in the Appendix.

|  |  | redox potential | org. acids | total residue | ignition residue | $Fe^{2+}$ | $NH_4^-$ | $NO_3^-$ | $SO_4^{2-}$ | $BOD_5$ | COD |
|---|---|---|---|---|---|---|---|---|---|---|---|
|  |  | 1 | 2 | 3 | 4 | 5 | 6 | 7 | 8 | 9 | 10 |
| redox potential | 1 | 1 | .78 | .44 | .03 | .41 | .72 | .46 | .48 | .79 | .84 |
| org. acids | 2 |  | 1 | .53 | .3 | .04 | .64 | .55 | .48 | .93 | .93 |
| total residue | 3 |  |  | 1 | .77 | .20 | .70 | .37 | .00 | .61 | .64 |
| ignition residue | 4 |  |  |  | 1 | .19 | .37 | .08 | .28 | .03 | .04 |
| $Fe^{2+}$ | 5 |  |  |  |  | 1 | .69 | .47 | .38 | .06 | .15 |
| $NH_4^-$ | 6 |  |  |  |  |  | 1 | .64 | .56 | .68 | .74 |
| $NO_3^-$ | 7 |  |  |  |  |  |  | 1 | .45 | .51 | .56 |
| $SO_4^{2-}$ | 8 |  |  |  |  |  |  |  | 1 | .42 | .46 |
| $BOD_5$ | 9 |  |  |  |  |  |  |  |  | 1 | .99 |
| COD | 10 |  |  |  |  |  |  |  |  |  | 1 |

**Fig. 17:** Correlation matrix of 10 parameters in seepage water from refuse dumps.

Group diagrams allow a direct comparison to be made e.g. of ground waters and their formation. Triangular or square diagrams are often used.

In the square diagram (Figure 18), each corner represents a component, and the percentages of cations and anions as milliequivalents of the total solution contents are plotted. In the present example, the four groups ($Na^+ + K^+$, $Ca^{2+} + Mg^{2+}$, $HCO_3^- + CO_3^{2-}$, $Cl^- + NO_3^- + SO_4^{2-}$) are plotted against each other. The patterns of points formed in this way can then be interpreted.

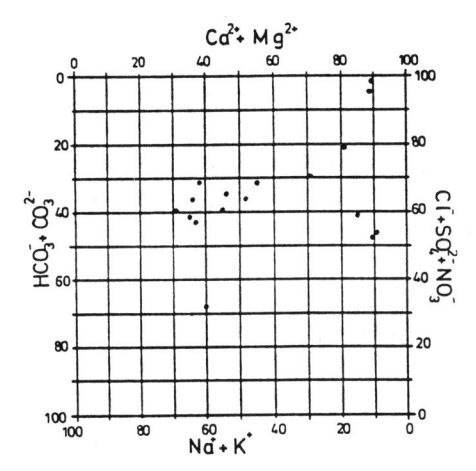

**Fig. 18:** Square diagram of water types

Calculation of the ion balance is of particular importance in the examination of groundwater, drinking water and surface water. Here, the sums of cation and anion equivalents must be equal. Normally, determination of sodium, potassium, ammonium, calcium, chloride, sulfate and nitrate concentrations together with the alkalinity to pH 4.3 is sufficient. First, the cation sum is obtained from the $H^+$ concentration (pH-value) and the ion equivalents of the other cations. The sum of anion equivalents is similarly obtained. Finally, the two sums of equivalents are listed and the analytical error calculated in percent. A computer programme (Program 2) for calculation of the ionic balance is listed in the Appendix.

## 3 Requirements for Analytical Methods

The specifications of the analytical methods for water, sludge and soil must of course depend on the related aims and expectations. The requirements must be well defined before embarking on extensive and costly measurement programs in order to attain an optimal cost-benefit relationship and a rapid supply of information.

The following questions should be answered before preparing the analytical instructions:

- for what purpose are the measurements required?
- which are the most important parameters enabling a general assessment to be made?
- which sampling points and measurement frequencies are absolutely necessary?
- how should the sampling and measuring programs be organized with respect to place and time?
- how should the obtained data be documented?
- what are the personnel requirements and costs?

These questions cannot be dealt with generally but require detailed examination for each individual assignment. For this reason, the following comments are only designed to provide some guidance in defining aims in connection with the benefits to be obtained.

In general, the following goals may be formulated for the examination of water, waste water and soil:

Water
- determination of the actual water quality (official water quality register),
- determination of local and temporal tendencies (monitoring program, forecasting models),
- determination of impurity sources,
- determination of suitability of water for a defined purpose,
- preparation for water technological processes.

Waste water
- measurements to allow estimation of potential damage caused by introduction into waterways,
- measurements to allow estimation of potential damage to canals and sewage treatment plants,
- preliminary work for the planning and operation of waste water treatment plants,
- testing of selected detrimental parameters for calculation of fees,
- initial examination before changing production processes.

Soil
- examination to allow estimation of soil quality for an intended use,
- optimization of fertilization programs,
- estimation of trends in soil changes (e.g. salting, acidity),
- examination in connection with ground water quality e.g. determination and release of contaminants.

Table 4 lists the test groups together with comments on their usage and limitations required for the assessment of water and soil.

**Table 4:** Test groups in the investigation of water and soil

| Test group | Comments |
|---|---|
| 1. Sensory Examination | essential for all samples; simple to perform (at collection or in the laboratory) |
| 2. Physico-chemical measurements | essential for all samples; simple to carry out (in part during collection) |
| 3. Group analyses | often carried out in the laboratory to provide information e.g. regarding the type of contamination |
| 4. Cations, anions and undissociated substances as major components | often carried out to characterize the general chemical composition; rapid methods are available (in laboratory) |
| 5. Inorganic trace analyses | less common; provide information- on e.g. inorganic contaminants; in some cases rapid tests are available (in laboratory) |
| 6. Organic trace analyses | less common; give information on organic contaminants; few rapid tests available (in laboratory) |
| 7. Biological parameters | allow conclusions to be drawn concerning the biological environment and hygienic properties; some tests can be made during sample collection |

## 3.1 Ground Water

Ground water is water which occupies subterranean permeable layers. Figure 19 is a schematical representation of ground water replenishment from rainfall: contact between percolating water and the earth, the unsaturated zone (partial filling of the porous voids), free ground water as an unconfined aquifer and a ground water-bearing stratum as a confined aquifer. The rate of percolation differs according to the type of ground material: the rate of flow in sands is 1 to 5 m per day, in gravel 6 to 10 m per day and in silts and clays (ground water dams) the rate can be as low as a few mm or cm per day.

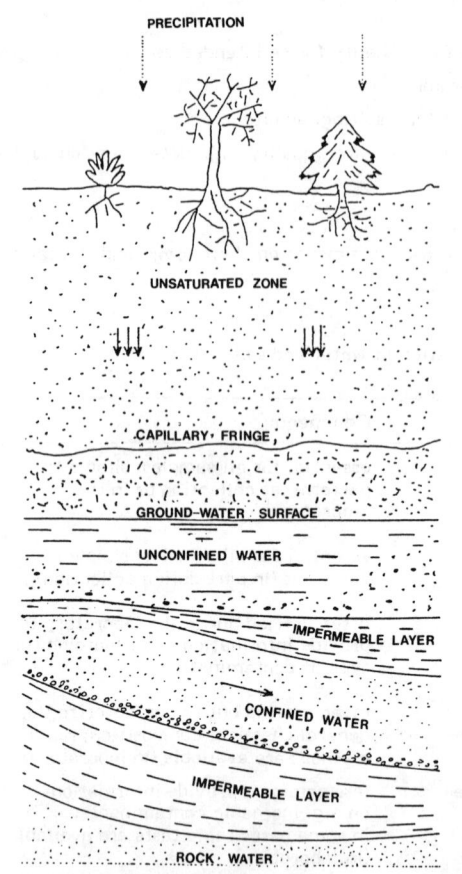

**Fig. 19:** Schematic representation of ground water replenishment and different types of ground water

The ground water quality i.e. type and amount of dissolved substances is determined by the properties of the soil, the rate of percolation as well as the time and depth of residence.

Depending on the geological and pedological properties, variations in water quality can be considerable. The following mass balance applies:

initial rock + atmospheric water → changed rock + solution

**Fig. 20:** Mass balance for natural systems

The diverse nature of the rocky subsoil and the various qualities of water thus formed allow several major categories to be determined. These groups may be readily recognized in a square diagram (Figure 21).

## 3.1 Ground Water

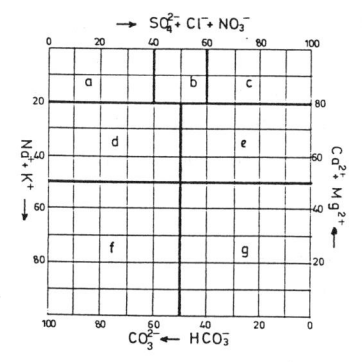

**Fig. 21:** Differentiation of important ground water types

Normal alkaline earth water:

a) mainly hydrogen carbonate (bicarbonate)
b) hydrogen carbonate-sulfate
c) mainly sulfate

Alkaline earth water with greater alkali content

d) mainly hydrogen carbonate
e) mainly sulfate

Alkaline waters

f) mainly hydrogen carbonate
g) mainly sulfate/chloride

The different ranges of the more important constituents depending on the type of ground water are shown in Table 5.

**Table 5:** Common concentration ranges of different ground water types

| parameter | magmatic rock | sandstone | carbonate rock | gypsum | rock salt |
|---|---|---|---|---|---|
| | mg/L | mg/L | mg/L | mg/L | mg/L |
| $Na^+$ | 5 - 15 | 3 - 30 | 2 - 100 | 10 - 40 | - 1000 |
| $K^+$ | 0.2 - 1.5 | 0.2 - 5 | - 1 | 5 - 10 | - 100 |
| $Ca^{2+}$ | 4 - 30 | 5 - 40 | 40 - 90 | - 100 | - 1000 |
| $Mg^{2+}$ | 2 - 6 | 0 - 30 | 10 - 50 | - 70 | - 1000 |
| $Fe^{2+}$ | - 3 | 0.1 - 5 | - 0.1 | - 0.1 | - 2 |
| $Cl^-$ | 3 - 30 | 5 - 20 | 5 - 15 | 10 - 50 | - 1000 |
| $NO_3^-$ | 0.5 - 5 | 0.5 - 10 | 1 - 20 | 10 - 40 | - 1000 |
| $HCO_3^-$ | 10 - 60 | 2 - 25 | 150 - 300 | 50 - 200 | - 1000 |
| $SO_4^{2-}$ | 1 - 20 | 10 - 30 | 5 - 50 | - 100 | - 1000 |
| $SiO_3$ | - 40 | 10 - 20 | 3 - 8 | 10 - 30 | - 30 |

Environmental factors (refuse dumps, waste water seepage, spillages) can cause contamination especially of the upper groundwater layers. These changes must be considered according to the potential usage. Ground water is used as crude water for drinking purposes, for industrial use, for feeding boilers and for irrigation. The analyses to be performed are therefore dependent on such requirements e.g. to determine the amount of purification necessary or the extent of underground contamination.

A list for the analysis of ground water is given in Table 6. These parameters may be varied according to the intended use of the water or extent of environmental contamination.

**Table 6:** List of parameters for the analysis of ground water

| Determination | General ground water examination | Tests for concrete corrosion | Tests of ground water contamination review | comprehensive |
|---|---|---|---|---|
| **1) Sensory examination** | | | | |
| odor, color, turbidity | x | x | x | x |
| **2) Physico-chemical** | | | | |
| temperature | x | | x | x |
| pH-value | x | x | x | x |
| electrical conductivity | x | | x | x |
| redox potential | x | | x | x |
| solids | | | | x |
| absorption at 254 nm | x | | | |
| **3) Group analysis** | | | | |
| residue | | | | x |
| ash | | | | x |
| oxidizability | x | x | x | x |
| dissolved organic carbon | | | | x |
| phenol index | | | | x |
| hardness (= sum of alkaline earths) | x | x | | |
| toxicity | | | | x |
| **4) Cations, anions, undissoc. substances** | | | | |
| sodium | x | | | x |
| potassium | x | | | x |
| ammonium | x | x | x | x |
| calcium | x | x | x | x |
| magnesium | x | x | x | x |
| iron (total) | x | | | x |
| iron-II | x | | | x |
| manganese | x | | | x |

**Table 6: continued**

| Determination | General ground water examination | Tests for concrete corrosion | Tests of ground water contamination review | comprehensive |
|---|---|---|---|---|
| hydrogen carbonate | x | x | x | x |
| chloride | x | x | x | x |
| nitrate | x | x | x | x |
| nitrite | | | | x |
| fluoride | x | | | x |
| cyanide | | | | x |
| sulfate | x | x | | x |
| sulfide | | x | | x |
| phosphate | x | | | x |
| silicic acid | x | | | x |
| metaboric acid | | | | x |
| oxygen | x | | x | x |
| carbon dioxide (free) | x | | | x |
| aggressive carbon dioxide | x | x | | |
| **5) Inorganic traces** | | | | |
| arsenic | | | | x |
| cadmium | | | | x |
| chromium | | | | x |
| copper | | | | x |
| lead | | | | x |
| mercury | | | | x |
| nickel | | | | x |
| zinc | | | | x |
| **6) Organic traces** | | | | |
| volatile halogenated hydrocarbons | | | | x |
| hydrocarbons | | | | x |
| involatile halogenated hydrocarbons | | | | x |
| polycyclic aromatic hydrocarbons | | | x | x |
| **7) Biological parameters** | | | | |
| colony count | | | | x |
| faecal indicators | | | | x |
| multi-cellular organisms | | | | x |

## 3.2 Surface Water

Surface water can be either flowing or still. It is especially endangered by environmental contamination so that special precautions must be taken for its protection depending on the envisaged use. The degree of pollution always parallels changes in the ecological situation. It is generally true that a minimum quality of surface water must be ensured in order to maintain the property of self-purification. This latter property can cause natural or foreign organic pollutants to be gradually degraded. In oxygen-rich zones, such degradable substances are broken down by the action of aerobic microorganisms. Oxygen is thereby consumed and its concentration in the water often falls. Should oxygen-free zones appear, certain anaerobic organisms thrive and rotting processes can start. The degradation of organic substances causes a biomass to be formed which is precipitaded as an organic sediment.

Physico-chemical tests provide important information on the actual state of the water, on possible pollutants or their origin and on the function of the self-purification processes. However, especially in the case of surface water, additional biological tests are necessary for a complete water quality assessment.

In order to be able to compare analytical results, the so-called "Saprobic Index" has been introduced. This is based on the fact that certain organisms known as indicator organisms are only found in water of sufficiently high quality. The biological results are normally complemented by further parameters such as the ammonium concentration, oxygen content and the biochemical oxygen demand in order to facilitate the evaluation. The types of organism present varies naturally from region to region so that Table 7 may only have limited validity.

In judging the quality of water and possible fluctuations thereof it should be taken into account that still water is more sensitive to pollutants than flowing water. Thus, eutrophy often occurs in lakes and reservoirs. Here, the phosphate concentration plays a decisive role since phosphate is a "minimum factor" leading to heavy algae production even in very small concentrations. Sources of phosphate include domestic waste water or the eroded surfaces of heavily fertilized ground. Also, an increased nitrogen content leads to increases in the production of algae and corresponding decreases in water quality.

The appearance of pathogenic organisms in surface water should be especially noted where the water is intended for use as drinking water or for bathing. In this connection, the presence of the following organisms should be determined:

- total coliform bacteria,
- faecal coliform bacteria,
- faecal streptococci,
- salmonella,
- intestinal viruses.

**Table 7:** Quality classification of surface waters

| Quality class | Saprobic level | composition of the biological environment | BOD$_5$ mg/L | NH$_4^+$-N mg/L | O$_2$ mg/L |
|---|---|---|---|---|---|
| I | Uncontaminated to very slight contamination | few bacteria; sparse population of gravel algae, red algae, moss; spawning ground of salmonides; mainly spring area | 1 | trace | 8 |
| II | Moderate contamination | slight organic contamination; very wide variety and density of algae, snails, small crabs and insect larvae; abundunce of water plants; good fish water | 2 - 6 | 0.3 | 6 |
| III | Heavy pollution | dense population of rod-shaped bacteria; few algae and higher plants; abundance of leeches, sponges, water lice; few fish | 7 - 13 | 0.5 | 2 |
| IV | Excessive pollution | only bacteria, fungi, protozoa; no higher organisms | 15 | several mg/L | 2 |

These pathogenic microorganisms can reach surface waters via domestic drains or waste water from hospitals or abattoirs. Since the optimum living temperature is 37 °C, such waters provide unfavorable conditions and the organisms do not normally multiply further. However, their infectiousness can be preserved by certain surviving forms. In normal biological sewage works, such pathogens can never be entirely removed so that waste water always adversely affects surface waters.

The physico-chemical test parameters for surface waters are very numerous. They originate from legal requirements or scientific and technical problems. More than 1000 parameters have been investigated in surface waters, and approximately 100 of these are frequently considered. However, the scope of the investigation can often be significantly reduced where knowledge of the type of contamination is available. Many parameters already named in connection with ground water monitoring are also employed for surface waters. In addition, the following are of special importance for surface waters: visual depth, oxygen content (O$_2$-saturation, O$_2$-deficit), general organic contamination (measured e.g. as biochemical and chemical oxygen demand), biological parameters and particular substance classes such as surfactants, mineral oils and pesticides.

For monitoring the minimum required quality of surface waters, the parameters which can be measured include:

temperature,
pH-value,
oxygen content,
ammonium content,
chemical oxygen demand (COD),
biochemical oxygen demand (BOD) without inhibition of nitrification,
total phosphorus content,
total iron content,
zinc content,
copper content,
chromium content,
nickel content.

The test program can be tailored to the main intended use of the water. Requirements and threshold values are given in Chapter 7. Such uses can include:

- commercial fish farming,
- angling,
- bathing,
- leisure and recreation,
- drinking water (indirect),
- drinking water (direct),
- irrigation.

## 3.3 Drinking Water

Drinking water is the most important material for human consumption and as such must be free of pathogenic organisms and possess no properties detrimental to health. Environmental contamination of drinking water can spread infectious diseases such as cholera, hepatitis, worms or typhoid. For this reason, continuous bacteriological examinations are highly recommended in order to prevent the danger of infection. In addition, harmful materials such as heavy metals, cyanides, phenols or pesticides can reach drinking water by various routes and laboratory tests are also necessary in these cases. In general, ground water or surface waters are used for the preparation of drinking water. Widely differing problems can arise during processing and distribution according to the source.

The crude water can be obtained from the following sources:

- groundwater,
- filtered surface water,
- spring water,
- groundwater enriched by surface water,
- groundwater enriched by purified waste water,

## 3.3 Drinking Water

- river water,
- natural lakes,
- reservoirs,
- desalinated sea water,
- rain water.

The scope of the examination must take into account the type of raw water. The control starts with the sources from which the raw water is drawn (sections 3.1 and 3.2) and continues through the storage reservoir and the various processing and distribution facilities, finally ending at the consumer. Figure 22 describes this procedure schematically.

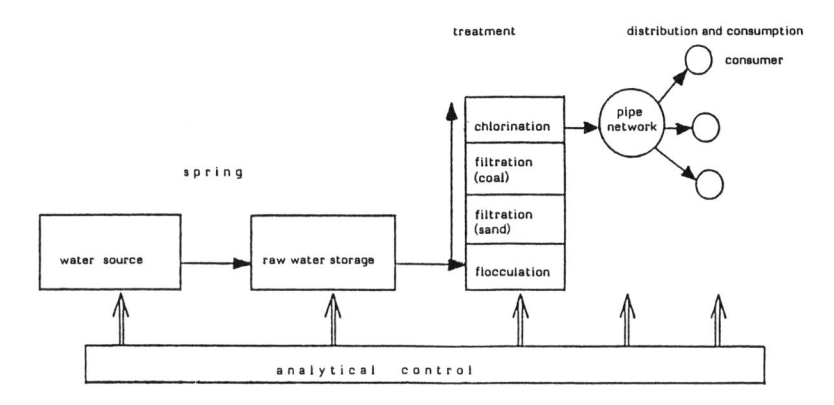

**Fig. 22:** Control areas in the treatment and distribution of drinking water

The minimum quality requirements of drinking water as well as of surface water designated as a drinking water source are given in Chapter 7. The scope of examination must be established depending on the relevant legal requirements and individual circumstances. For the day-to-day control of the water processing, a more limited or modified test scope may be employed. Group determinations instead of single analyses may be introduced to improve efficiency or increase the sample turnover in those cases where no safety reduction would be caused.

The rapid availability of analytical results is of great importance in the examination of drinking water so that the steps necessary to improve the water quality (e.g. deacidification by mixing of various raw waters, sterilization through changing the raw water quality) can be taken without delay. Compared with the examination of other water types this requires:

- a well functioning laboratory organization,
- well proven, rapid and reliable test method,
- practized documentation procedures and data interpretation,
- immediate application of results on a technological scale.

## 3.4 Waste Water

The term waste water is used to refer to water which has had its composition and character changed by intensive human usage. Waste waters can differ greatly so that neither a single classification nor a standard examination procedure are possible.

Waste water can significantly pollute ground and surface waters. The greatest contamination is caused by the drainage system from towns, industrial sites and agriculture. Damage is caused by waste waters which reduce the oxygen content of other waters through dissolved ingredients which use up oxygen (e.g. organic compounds, ammonium, sulfites). In addition, effects of fertilizer (eutrophication) can be introduced by the presence of nutrients such as nitrogen and phosphorus.

A special characteristic of many industrial waste waters is the content of those components which can hinder the process of self purification. The components may be grouped according to the negative effects which they cause:

Toxic substances can cause acute or chronic poisoning of water organisms.

Interfering substances can cause undesirable odor, taste, color, cloudiness as well as technical problems during processing, distribution and use.

Degradable substances can reduce the oxygen content.

Nutrients cause the eutrophication of still or slow moving waters.

The damage caused by industrial waste waters to the primary source affects all further usage and especially the suitability as a drinking water supply, for fisheries, agriculture and animal farming. Before introduction into the primary source, the waste water should be treated so that any potential damage is reduced to an extent that the balance of properties relative to the end applications of the water are not adversely affected. For this purpose, the waste water control examination plays a very important role.

Not all primary sources have the same intended usage and hence the various requirements associated with these uses give rise to a variety of properties and standards which have to be met and controlled. Thus water quality standards may be defined for the various categories of usage. However, even within the industrialized nations, no universal standards have yet been adopted for this purpose.

A general classification of waste water is only possible in the domestic case. Table 8 gives selected parameters for a heavy, medium and weak contamination.

## 3.4 Waste Water

**Table 8:** Average parameter concentrations in domestic waste waters with different degrees of contamination

|  | Contamination |  |  |
|---|---|---|---|
| Parameter | heavy mg/L | medium mg/L | weak mg/L |
| Total solids | 1000 | 500 | 200 |
| Sedimentable matter (mL/L) | 12 | 8 | 4 |
| Biochemical oxygen demand (BOD) | 300 | 200 | 100 |
| Chemical oxygen demand (COD) | 800 | 600 | 400 |
| Total nitrogen | 85 | 50 | 25 |
| Ammonia - N | 30 | 30 | 15 |
| Chloride | 175 | 100 | 15 |
| Alkalinity (as $CaCO_3$) | 200 | 100 | 50 |
| Oils and fats | 40 | 20 | 0 |

The introduction of waste water components into sewage plants can inhibit biological processes. Such damage may be expected to occur where the following concentrations are found (Table 9):

**Table 9:** Concentration ranges of toxic substances causing inhibition of biological processes in sewage works

| Toxic Substance | Concentration range (mg/L) |
|---|---|
| Copper | 1 to 3 |
| Chromium-III | 10 to 20 |
| Chromium-VI | 2 to 10 |
| Cadmium | 3 to 10 |
| Zinc | 3 to 20 |
| Nickel | 2 to 10 |
| Cobalt | 2 to 15 |
| Cyanide | 0.3 to 2 |
| Hydrogen sulfide | 5 to 30 |

As already mentioned, the examination of waste waters can differ greatly. The following examinations should be considered (Table 10).

**Table 10:** List of parameters for the analysis of waste waters

| Parameter | Short analyses | Analyses to determine oxygen consumption | Comprehensive analyses |
|---|---|---|---|
| Color, odor | x | | x |
| pH-value | x | x | x |
| Electrical conductivity | x | | x |
| $KMnO_4$-consumption | x | x | |
| Chemical oxygen demand (COD) | x | x | x |
| Biochemical oxygen demand (BOD) | x | x | x |
| Total organic carbon (TOC) | | | x |
| Dissolved organic carbon (DOC) | | | x |
| Total nitrogen | | x | x |
| Ammonium, nitrite, nitrate | x | x | x |
| Total phosphorus | x | x | x |
| Fluoride | | | x |
| Sulfate | x | | x |
| Sulfide, sulfite | | x | x |
| Chloride | x | | x |
| Hydrogen carbonate | x | | x |
| Sodium, potassium | | | x |
| Calcium, magnesium | x | | x |
| Iron, manganese | x | | x |
| Heavy metals (As, Pb, Cu, Ag, Zn, Cr, Cd, Hg, Ni) | | | x |
| Digestibility | x | x | x |
| Fish toxicity | | | x |
| Toxicity to microorganisms | | | x |
| Total phenols, volatile phenols, organically bound chlorine, cyanides | | | x |
| Detergents | | | x |
| Pesticides | | x | |

## 3.5 Soil

Soil refers to the uppermost part of the earth's surface which has been formed under the

3.5 Soil    43

influence of weathering. The soil consists of mineral and organic substances (humus) and generally contains water, air and living organisms.

Soils represent the supporting environment for the higher plants and together they form an ecological system. The suitability of soil for this purpose are mainly dependent on the space available for root growth and the availability of water, air, heat and, nutrients. These determine the fertility of the soil. The soil/plant ecological system is illustrated schematically in Figure 23.

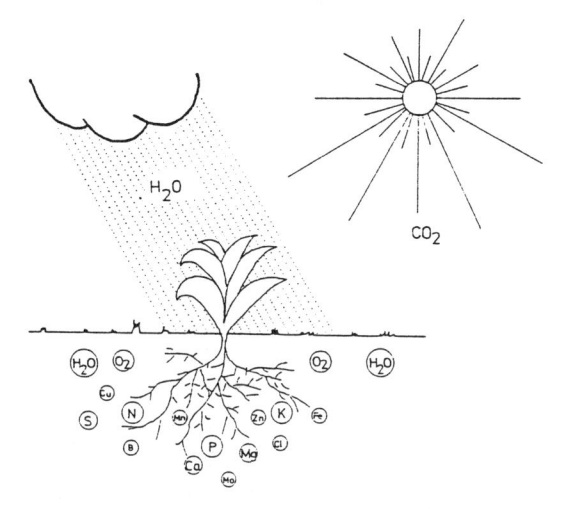

**Fig. 23:** Schematic representation of the soil/plant ecological system

The space available for root growth is generally determined by the solum depth. However, loose stones, incorporated into the soil mean that the roots can penetrate deeper. The degree of root formation strongly depends on the soil consistency as well as factors such as density variations, pore content, ground water level and distribution, salt content, pH-value and redox potential.

The degree of resistance to root penetration is defined as follows:

**Table 11:** Root penetration in soil

| Depth | Root Penetration |
|---|---|
| less than 10 cm | very shallow |
| 10 to 25 cm | shallow |
| 25 to 50 cm | medium |
| 50 to 100 cm | deep |
| over 100 cm | very deep |

The examination of soil can achieve the following:

- determination of the nutrient availability
- determination of the proportion of those substances having a damaging effect on the ground water.
- determination of corrosive properties for pipe-lines
- determination of salinization
- test for filtering efficiency

The amount of mineral nutrients available in the substrates is of decisive importance for the supply to plants and they must be determined using suitable methods. Thus, the necessary amount of fertilizer to be added can be calculated.

The availability of nutrients is summarized in Table 12:

**Table 12:** Forms of nutrient bound in soil

| Bound form | Availability | To be determined |
| --- | --- | --- |
| Unbound dissolved in the soil | very ready | water soluble nutrients |
| Partly bound to exchangers | ready | exchangeable nutrients (incl. water soluble) |
| Immobile, readily mobilized | moderate | exchangeable nutrient (incl. water soluble) |
| Immobile, not readily mobilized | very slight | total reserve of nutrients |

However, the available proportions of the single nutrients depend on the clay content, humus content, soil humidity, pH-value and redox potential. The mobilized amounts actually depend on the climatic conditions and soil horizon.

The assessment of inorganic contaminants in soil is difficult since the natural concentrations e.g. of heavy metals can be high depending on the mineralogical situation. The addition of nutrients during fertilization cannot be considered to be a contamination just as the removal of nutrients by plants cannot be considered to be a decontamination. The evaluation of contamination can be considered from the point of view of profitability or the efficiency of the soil when regarded as a water filter. Here, the important aspects are:

- the economic importance of the soil as a substrate for cultivated plants,
- the filter capacity of the soil for protecting the ground water.

The experimental determination of the mobile share of heavy metals involves practical problems. In order to measure the share of heavy metals actually available to the plant a method of extraction which allows a comparison with the heavy metal uptake of the plant roots must be used. Strong acids are of course not suitable as extractants and this is also true of very strong complexing agents. Normally, an elution method with the aid of pyrophosphate or dithionite-citrate is employed. Extracts with ammonium hydroxide are also suitable for determining the mobile fraction of trace elements. The extractability of most heavy metals from soil is greater than that of most common elements (phosphorus: 1%, aluminium: 2 to 3%, sodium: 0.1 to 0.5%). However, it is known that heavy metals are more strongly bound in humus material than the elements mentioned above.

The tolerable limits of heavy metals in soil are of importance. Table 13 provides toxicity thresholds above which concentrations an impairment of plant growth would be expected. The limits vary greatly from plant to plant.

**Table 13:** Toxicity ranges of various heavy metals in soil

| Element | Toxicity threshold mg/kg |
|---|---|
| Copper | 200 - 400 |
| Zinc | 500 - 5000 |
| Cadmium | 10 - 175 |
| Nickel | 200 - 2000 |
| Lead | 500 - 1500 |
| Chromium | 500 - 1500 |
| Mercury | 10 - 1000 |

The contamination of soil with organic substances such as pesticides can be similarly considered. Only a part of the material will be applied directly on to or in the soil. However, on spraying, it is impossible to avoid soil contact caused either by material dripping from the plants themselves, or following subsequent rainfall. The absolute amounts which reach the soil via these routes are relatively small. The fate of pesticides in soil is illustrated in Fig. 24.

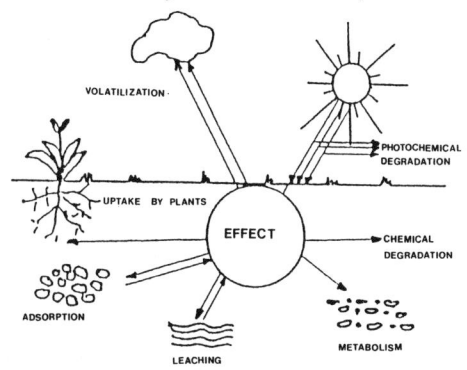

**Fig. 24:** Fate of pesticides in soils

These substances are subjected to a variety of biotic and abiotic influences which determine their properties in the soil, their transformation and eventually their complete mineralization. Humus surface soil layers adsorb a substantial proportion of the pesticides which reach the soil. Their further penetration depends on content of clay minerals, nitrogen containing compounds, iron and aluminium oxides and the amount of macropores in the soil.

Examination of the corrosive properties of the soil towards metal pipe-lines as well as the testing for salinization, especially in arid areas form a special area of investigation which will not be discussed here in depth (see sections 6.2.3.6 and 7.5).

In Table 14 the possible test programs depending on the experimental objectives are listed for soil examinations.

**Table 14:** List of parameters for soil analysis

|  | nutrient content | Assessment of toxic substances | corrosion of metal pipes | salinization |
|---|---|---|---|---|
| Particle size distribution | x |  | x | x |
| Water content | x |  | x | x |
| pH-value | x | x | x |  |
| Redox potential |  | x | x |  |
| Electrical conductivity | x | x | x | x |
| Acidity/alkalinity | x |  | x | x |
| Organic carbon | x |  | x |  |
| Sodium | x |  |  | x |
| Potassium | x |  |  | x |
| Calcium | x |  | x | x |
| Magnesium | x |  | x | x |
| Manganese | x |  |  |  |
| Copper | x | x |  |  |
| Zinc | x | x |  |  |
| Cr, Ni, Cd, Hg |  | x |  |  |
| Boron | x | x |  |  |
| Molybdenum | x |  |  |  |
| Total N, organic nitrogen | x |  |  |  |
| Ammonium | x |  |  |  |
| Nitrate | x |  |  |  |
| Total phosphorus | x |  |  |  |
| Chloride |  |  | x | x |
| Sulfate |  |  | x | x |

**Table 14:** continued

|  | Assessment of | | | |
| --- | --- | --- | --- | --- |
|  | nutrient content | toxic substances | corrosion of metal pipes | salinization |
| Carbonate, hydrogen carbonate |  |  |  | x |
| Sulfide |  |  | x |  |
| Pesticides |  | x |  |  |
| SAR-value (see section 6.2.3.6) |  |  |  | x |

# 4 Organization of Sample Collection Program and Sampling Techniques

## 4.1 General

Test results are dependent upon the samples which have been taken from a whole population. Therefore, very special attention has to be paid to the organization of the collection and the sampling techniques themselves. The main aim is to obtain samples which are representative and valid for the population. This means that the samples must be collected and stored in such a way that the parameters measured in the sample themselves have the same values as those in the sampled water, waste water or soil. Furthermore, the sampling place and time must be chosen in such a way that the analytical results reflect the temporal or local variations during the period of investigation.

Every sample collection depends to a certain extent on chance and is therefore burdened with an inherent error. The smaller the sample, the less representative it is of the population. In addition, the information contained in the result of a random sample depends on the variation of the particular parameter measured.

In order to generalize empirical results, the size of the random sample error must be known. The random sample error describes the difference between the value of a random sample (e.g. arithmetic mean) and the "real" value for the whole population. The size of random sample error depends on the sample population. Above a certain number of samples, the random sample error becomes so small that an increase in the sample population cannot be justified.

The following goals can be pursued by taking samples:

- quality control,
- making forecasts,
- determination of the extent of damage.

The majority of sample collections are for the purpose of quality control, often in connection with official requirements. In making forecasts, data is obtained in order to solve subsequent problems or to recognize contamination trends, e.g. before construction of a sewage works when a forecast on the future influence of waste water on surface water is required. In cases of damage investigation, questions of cause and extent are often of primary importance.

Representative and valid samples of waters and waste waters can be obtained in several ways. A preliminary choice of the sampling method may be made according to the following scheme:

**Table 15:** Preliminary choice of sampling method

| Concentration fluctuation | small flow variation | large flow variation |
|---|---|---|
| small | random sample | random sample |
| large | time proportional pooled sample | volume proportional pooled sample |

Figure 25 illustrates the various types of sample collection.

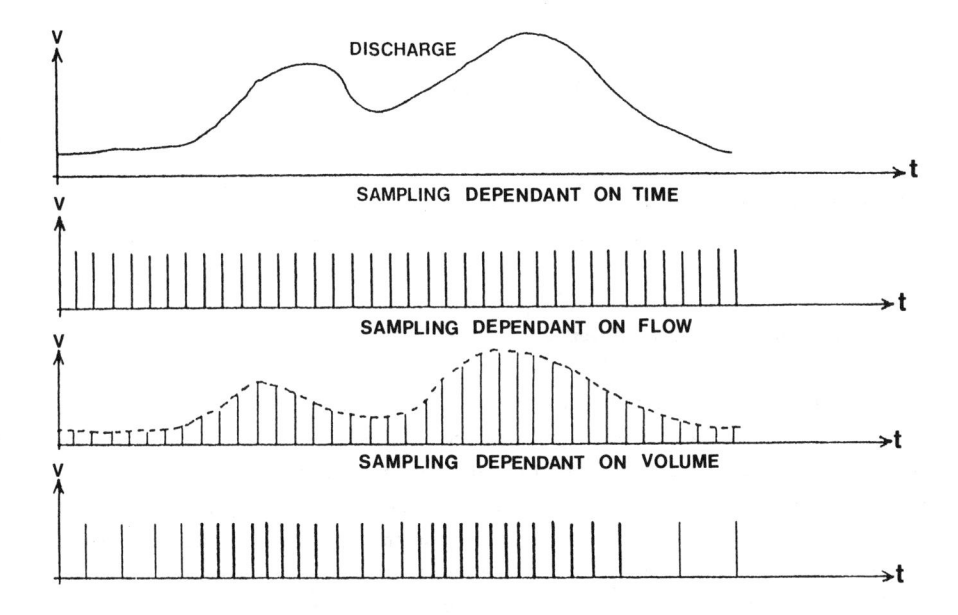

**Fig. 25:** Possible water sampling techniques

Random samples are mostly collected manually from pipes, reservoirs, rivers, streams or drains. However, random samples can also be taken automatically. Each sample provides data on waste water quality at the collection time only. Pooled samples can be manually prepared by mixing single samples or can be obtained from an automatic sampler according to times or quantities. Such samples enable conclusions to be drawn about the water quality over longer periods. However, in most cases, peak concentrations cannot be identified in such pooled samples. Time proportional sampling consists of taking constant volume samples at identical time intervals and pooling them. In quantity-proportional sampling, volumes proportional to flow are collected at constant time intervals. Where the flow is continuously monitored, an almost quantity-proportional pooled sample can be prepared (even manually) by mixing the relevant amounts from various single samples.

**Table 16:** Example of a quantity-proportional mixed sample

| Time | Flow (m$^3$/s) | Proportion of random sample in pooled sample (L) |
|---|---|---|
| 8  | 1   | 0.1  |
| 9  | 2   | 0.2  |
| 10 | 1.5 | 0.15 |
| 11 | 4   | 0.4  |
| 12 | 0.5 | 0.05 |
| 13 | 1   | 0.1  |
| 14 | 2   | 0.2  |
| 15 | 1.5 | 0.15 |
| 16 | 0.5 | 0.05 |
|    | Total: | 1.40 |

Also in the case of soil sampling, the requirement of representativeness must be satisfied. Random samples and average samples may be taken from a larger area. Details are given in section 4.6.5.

## 4.2 Organization of Sampling Networks

In those cases where no reference can be made to an existing measurement network for the examination of water, waste water or soil (e.g. new edition of an official pollution register), such a network should be planned. The preliminary plans and tests for this purpose should be carried out with special care as the network may have to be in use for many years and experience shows that subsequent modifications lead to difficulties in comparing analytical data. The major goal should be to obtain samples which are representative of the whole. Since complex systems such as larger catchment areas or the canal systems of larger cities normally consist of single sub-systems which differ in quality, attempts must be made to identify such sub-systems and to characterize them by separate examination.

Preliminary examinations are firstly carried out by application of the following techniques.

- topographic and specialized maps (ground maps, geological maps),
- aerial photographs,
- waterway plans, drainage plans.

In those cases where such accessories are not sufficiently available, provisional maps and plans must be drafted after local inspection. Further information is obtained by simple tests carried out in the field. It is important that a collection site be chosen not because of convenient access but because a representative sample is obtained.

## 4.2 Organization of Sampling Networks

The organization of measurement networks in waters and drainage systems is considered here so that the following description of systems of decreasing scale should be of some assistance. In the case of ground water measurement networks, the assistance of hydrogeologists is to be included in the plan. Figure 26 shows the catchment area of a river.

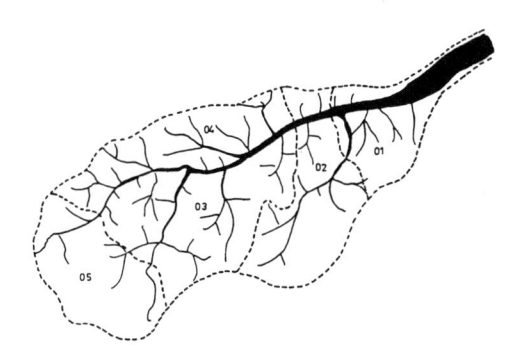

**Fig. 26:** Example of a catchment area

The whole area can be divided, e.g. into 5 single areas each of which has a measurement point marked on it. Depending on the particular assignment, further distinctions can be made. Critical areas in larger waterway systems lie beneath the junctions with strongly polluted tributaries and waste water inflow points. In addition, less contaminated places in the upper reaches should also be included in order to determine the natural properties of the water. A further aspect is to include measurement points on both sides of political borders in order to simplify the solution of potential disputes. Apart from industrial areas with a necessity of having many measurement points, one sampling point per 100 to 200 km$^2$ generally suffices.

In Figure 27, part of the whole system is shown schematically. Measurement points at which serious changes in water quality might be expected are shown as circles. Populated areas have been shaded and waste water entry points (sewage, industry) are marked by arrows.

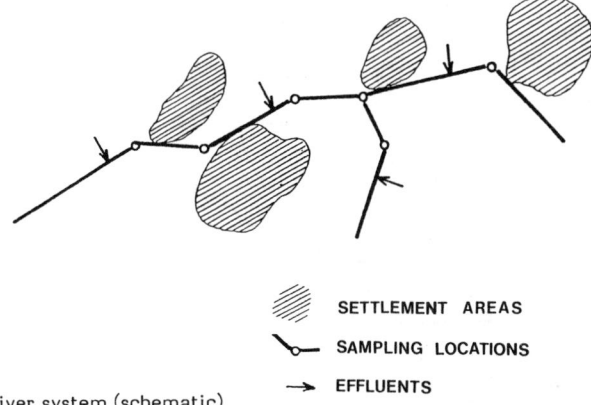

SETTLEMENT AREAS
SAMPLING LOCATIONS
EFFLUENTS

**Fig. 27:** Part of river system (schematic)

In additon to the major measurement points, regular checks can also be made at these entry points.

Figure 28 is a schematic representation of a network for waste water disposal in towns.

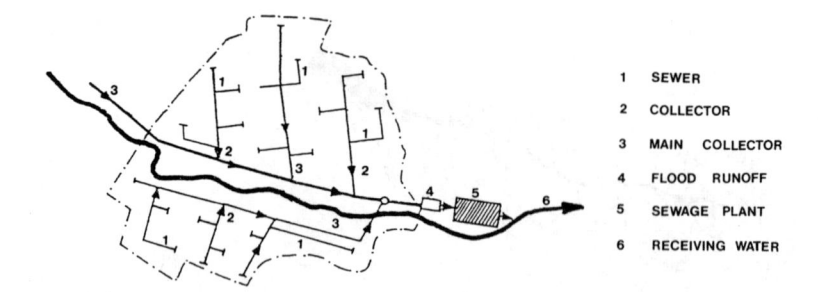

| | |
|---|---|
| 1 | SEWER |
| 2 | COLLECTOR |
| 3 | MAIN COLLECTOR |
| 4 | FLOOD RUNOFF |
| 5 | SEWAGE PLANT |
| 6 | RECEIVING WATER |

**Fig. 28:** Scheme of a sewage network

It contains the street canals, collectors, main collector, collection shaft, overflow chamber and overflow canal as well as the sewage plant with outlet into the waterway. The shaded areas represent industrial areas. Measurement points in the canal system should be set up where single plants deliver waste water to the network, where the major collectors meet and where sewage plants have their outlets. On suspicion of introduction of poisonous materials, the guilty party in the network can thereby be determined.

The precise sampling point in larger waterways or canals must be located such that entering waters above the sampling points are completely mixed. Should this not be the case, two sampling points can be positioned, one on each side of the waterway. Figure 29 is designed to show that the mixing of waters from tributaries or waste water inlets is very slow in the absence of turbulent currents.

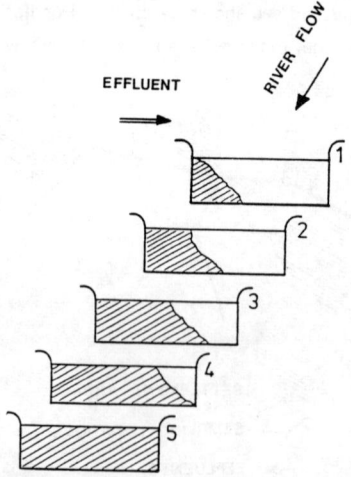

**Fig. 29:** Scheme of effluent mixing in flowing water systems

## 4.2 Organization of Sampling Networks

In such a case, a permanent sampling point should not be placed before position 5.

In still water systems (lakes, reservoirs) with irregular shapes, a considerable heterogeneity in water quality in the horizontal plane is possible. It is therefore necessary to arrange several sampling points which under certain circumstances may be reduced in number at a later date. Such a strong heterogeneity can also occur in the vertical plane so that sampling must be carried out at different depths e.g. near the surface, near eventual temperature layers and near the bottom (influence of sediments).

In addition to the choice of a representative spatial network for the sampling procedure, the choice of a suitable sampling timetable is also of considerable importance. Where great concentration fluctuations are to be expected, the frequency of sampling must be increased. In waters or waste waters with only slight variations, sampling at intervals of even several months can be sufficient for an assessment. Where cyclic variations in days, months or years are known, the sampling interval must be varied in order to avoid repeated measurement of a high or low value (Figure 30).

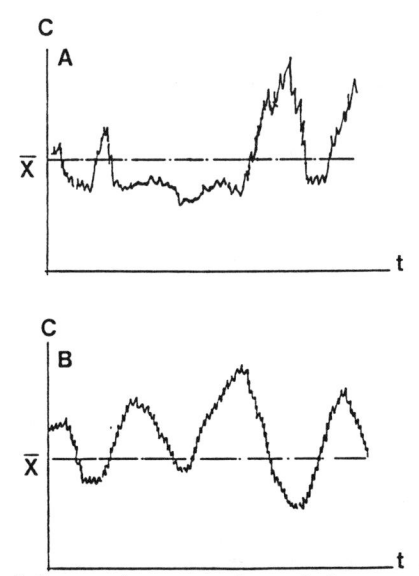

**Fig. 30:** Influence of concentration variations on values predicted from random samples (A-random; B-cyclic)

The application of random numbers, e.g. lot drawing has proved useful in maintaining random sampling times.

After a longer test period, the more important statistical parameters such as mean, variations and maxima should be determined. In addition, the occurrence of cyclic concentrations fluctuations may be detected graphically. Such evaluations can often allow the sampling frequency to be significantly decreased without loss of important information.

## 4.3 Determination of Water Quantity

Practical experience has shown that the flow and quantity determination of water and waste water is a relatively difficult task with problems related to the fact that perfect measuring systems and instruments for all applications do not exist.

Flow measurement only during sampling of ground water and drinking water can be easily carried out. In both cases, the measurement is made either with a stop-watch and graduated container or by reading a water gauge within a definite time interval.

In open canals or streams, the measurement has to be made with a large degree of technical and mathematical effort. More exact determinations normally employ systems relating to Bernoullis flow equation. Here, some well proved and for many applications sufficient methods will be described:

Open Canals
The method of Manning-Strickler may be employed for open canals. The canal gradients, wall roughness and the hydraulic circumference must be known. Where these three values are given, the flowing water quantity can be simply determined by measuring the water level.

$$v = k \cdot R^{2/3} \cdot J^{1/2} \quad (m/s)$$

| | |
|---|---|
| v | velocity (m/s) |
| R | hydraulic radius $\frac{F}{U}$ (m) |
| F | cross section (m$^2$) |
| U | wetted perimeter (m) |
| k | roughness (m$^{1/3}$/s) |
| J | gradient (m/m) |

The following standard k values are appropriate:

| | |
|---|---|
| 90 - 135 | for new asbestos cement pipes |
| 85 | for new stoneware and steel pipes |
| 65 - 75 | for encrusted pipes |

For right angled canals, R has the value:

$$R = \frac{\text{Width} \cdot \text{Water Depth}}{(\text{Width} + 2 \cdot \text{Water Depth})}$$

The flow is then:

$$Q = v \cdot F \quad (m^3/s)$$

## 4.3 Determination of Water Quality

### Open canals with irregular cross section (streams, rivers)

#### a) Float

The velocity of waterways can be determined as follows:

A float (e.g. a cork) is thrown into the middle of the flowing waterway and the time taken to cover a certain distance (Fig. 31) is measured

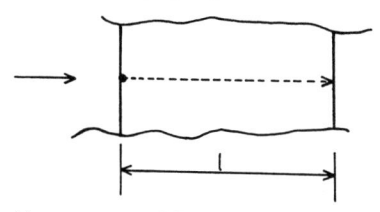

**Fig. 31:** Measurement of flow velocity by floats

Therefore:

$$v = \frac{l}{t} \quad (m/s)$$

$$Q = \frac{l}{t} \cdot F \cdot c \quad (m^3/s)$$

l     measured distance (m)

t     measured time (s)

F   = cross section of flow ($m^2$)

c   = coefficient

The flow cross section can be graphically determined or estimated by measuring the depth at various points.

For waterways lacking dense undergrowth on the banks and without coarse pebbles, k can be assumed to have a value of from 0.8 to 0.9; otherwise a value of 0.5 to 0.8 is taken.

#### b) Measuring by hydrometric propeller

The propeller method is employed in those cases where larger flow cross sections are encountered and the requirements are more stringent. Calibrated flow meters are employed (Figure 32).

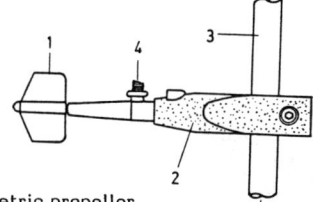

**Fig. 32:** Hydrometric propeller
1 - blades, 2 - instrument body with impulse transfer
3 - supporting rod, 4 - electrical contact

The meter is dipped into the water on vertical graduated plumb lines and the velocity measured at these points. Less stringent requirements may be satisfied by measuring the velocity at 0.4 x depth. The procedure and evaluation of results are carried out as shown in Figure 33.

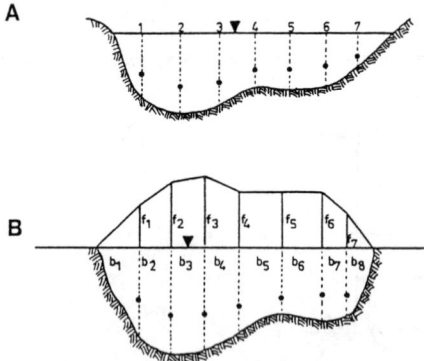

**Fig. 33:** Calculation of flow results using point measurements (Herrmann, 1977)

A - cross section of waterway showing plumb line measurement points

B - graphic determination of flow

Measurements are made at the points shown at 0.4 x depth. The velocity is then multiplied by the water depth and plotted as $f_i$. The $f_i$ points are then joined as shown to form a polygon and the sum of the single areas obtained from $f_i$ and the single widths $b_i$. The flow is then calculated as

$$Q = 0.5 \cdot \sum_{i=1}^{k} ( f_i \cdot b_i + f_k \cdot b_{k+1})$$

c) Measuring weir
Measuring weirs are suitable for irregular flows not exceeding 1 m$^3$/s. Triangular measuring weirs are often employed. The construction is illustrated in Figure 34.

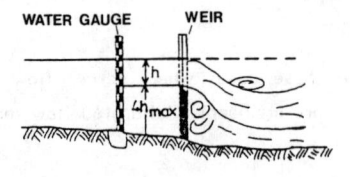

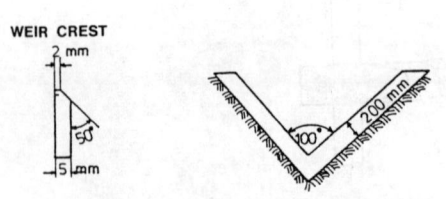

**Fig. 34:** Triangular weir according to THOMPSON

If the supply channel is triangular for 2 m like the weir itself and has the dimensions shown in the figure, Q is calculated as follows:

$$Q = \mu \cdot h^{5/2}$$

h = overfall height

$\mu$ = 1.46 (at the dimensions shown)

d) Salt mixing procedure

This technique is readily applicable in turbulent waters with rapid mixing. A salt solution of known concentration is required together with an electrical conductivity meter. The salt solution is allowed to flow under defined conditions into the waterway. The flow is then calculated as:

$$Q = A \cdot (B - C) \cdot (C + D)^{-1}$$

A   flow of salt solution

B   conductivity of salt solution

C   conductivity of river water with salt solution

D   conductivity of river water without salt solution

## 4.4 Sampling Devices

In general, the following requirements must be satisfied by devices which come into contact with the samples collected:

- robust, suitable for use in the field, easily transported, handled and cleaned,
- the material from which the device is constructed should cause no alteration in the sampling properties.

Depending on the type of sample collection, simple apparatus or specialized equipment (pumps, automatic samplers) are employed. Further auxiliary equipment such as cool boxes, tubes, ropes, cables, plastic bags, spades or gas burners to ensure sterile bottling of samples for bacteriological tests are not further mentioned here.

The choice of sample container is of prime importance. In most cases, containers of polythene or glass are employed. Materials which are not suitable for use with water containing certain substances should be noted. Samples having nonpolar organic contaminants (oils, pesticides), should not be placed in plastic containers whereas glass containers are unsuitable for use with water in which low concentrations of sodium, potassium, boron or silicic acid are to be determined. Both types of container are suitable for use with higher concentrations of inorganics.

Used containers must be thoroughly cleaned after use. Especially after longer storage periods of contaminated samples, solids can settle on the container walls and be strongly adsorbed. Cleaning is then normally carried out first mechanically and then with chromic acid. The latter must not be used with plastic containers. In these cases diluted hydrochloric acid is employed.

Stubborn impurities in plastic bottles cannot be completely removed. Such bottles are discarded. It should be remembered that use of household-detergents can cause surfactants and phosphates to be released from the surfaces even after thorough rinsing with water.

Sample containers for bacteriological examinations must be of glass and sterilized before use by heating for some time at 180 °C together with their stoppers. The bottle necks are protected by aluminium foil.

The sample collecting itself may be performed manually with a ladle or automatically with suitable instruments. In the case of ground water these are mainly dip pumps and for surface and waste water pumps coupled with a distribution valve. Soil samples may be taken with a spade or drill. Such specialized equipment is described below:

Groundwater
Ladles are the simplest devices for the collection of water samples. They are mainly employed for preliminary examination purposes or in those cases where the ground water mains are well mixed. They consist of cylindrical containers which are thoroughly washed through with water when lowered into the spring. At the sampling depth a valve is closed by means of some mechanism (e.g. activation through drop weight). The devices are therefore ideal for the defined sampling at any depths. Attention must be paid to the tightness of the seals. Such ladles are available in various diameters so that narrow springs can also be sampled.

Spring sampling is mainly carried out using pumping systems. In the case of suction pumps, the sampling depth is limited to 9 m since in deeper waters the column of water breaks in the sampling tube. A valve at the tube end is essential below 3 m.

Suction pumps can be powered by electrical or petrol motors. At smaller pumping depths the delivery capability of the pump can be between 40 and 180 L/min. For sampling below 9 m, an underwater pump is necessary. Such a submergable pump can be a centrifugal pump connected to a submergible motor. The smallest pumps have a diameter of 95 mm and can therefore only be employed in artesian wells of over 100 mm diameter. In the case of "bent" artesian wells of 100 mm diameter the pump can easily be lost. The pump is let into the well connected with a safety cable, a power cable and the sampling tube. A generator is used to deliver power where no mains supply is available. The conveying-capacity is found in the manufacturers description. Such submergible pumps are employed in well mixed ground wells. They should not be employed in cases of limited water supply since the water level can rapidly sink causing the pump to take in air. Further pump types are piston displacement, compressed air or deep suction pumps (water jet principle).

Pumps and tubes should be cleaned and dried after use to prevent corrosion of metal parts and build-up of microorganisms on plastic components.

Surface Water and Waste Water

Manual devices for the sampling of water from or near the surface typically include ladles; sealable sampling devices are employed to sample deeper layers. Automatic sampling devices have the following parts:

- delivery systems,
- control mechanism,
- sample apportionment and measurement,
- sample storage.

Delivery is accomplished by means of submergible pumps, peristaltic pumps or eccentric screw pumps. A system with a free-falling water device is shown schematically in Figure 35.

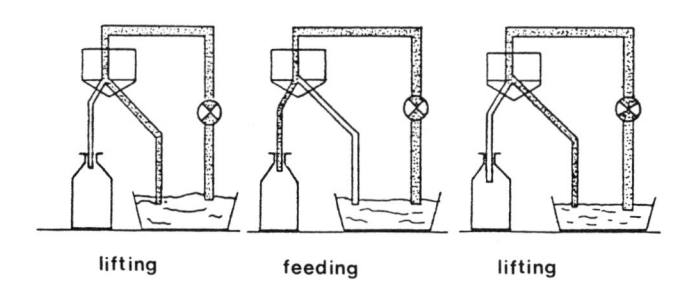

lifting          feeding          lifting

**Fig. 35:**   Schematic representation of an automatic free-fall sampler

In this way, a sample is diverted according to an impulse circuit (time or quantity proportional) from a continuous water stream. This type of sampler is easy to use and clean. A cooling unit should be incorporated for longer sampling periods.

Figure 36 shows the sampling phases of a system with free falling water device coupled with a constant sample dosing device.

In this way a constant sample volume is always obtained. Strongly contaminated waste waters can lead to blockage of the dosage container.

Short sealed pipes, high delivery velocities and the elimination of light must be ensured in order to minimize possible changes in the nature of the sample. Regular cleaning and maintainance should prevent deposit formation and ensure the reliability of the system.

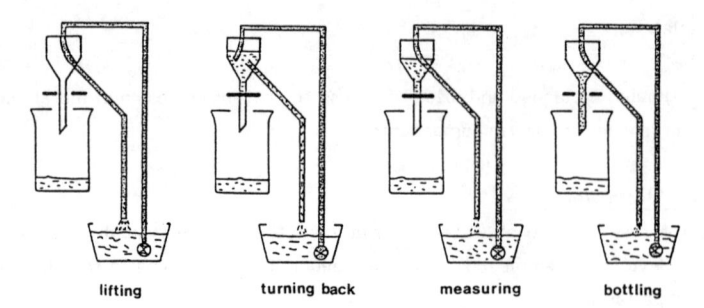

**Fig. 36:** Schematic representation of an automatic free-fall sampler

Soil

Sampling of soil for nutrient or salt testing generally take place at depths not exceeding 50 cm. Thus, a spade or simple boring device is normally sufficient. The latter can be a simple drill or a special drill head incorporating cut out walls which is forced into the ground with the aid of a heavy mallet and removed by turning. The soil profile can also be described in the drill grooves.

### 4.5 Preservation, Transport and Storage of Samples

The contents of water samples can alter at very different rates. As only a few parameters can be measured during collection, a preliminary treatment or stabilization is essential in many cases. These steps allow tests to be carried out even after long periods of time have elapsed. However, many ingredients are so stable that no special precautions for transport and storage are necessary. These are mainly the inorganic components. Waters which are absolutely free of organics and microorganisms show no microbiological degradation processes. However, physico-chemical changes are possible. Where large amounts of organic materials are present and the conditions suitable for growth of microorganisms, the water contents can change in a very short time. Preservation is highly recommended in such cases and is defined as the blocking or delay of biochemical processes, the aim being to prevent more than a 10 % change in contents between the original and the preserved sample.

Physical preservation of water samples by cooling is very important during transportation especially at high ambient temperatures since temperature increases cause an increase in the rate of biochemical processes. The following changes are possible in water samples:

- oxidation of components via dissolved oxygen (e.g. $Fe^{2+}$, $S^{2-}$)
- precipitation and co-precipitation of inorganics through milieu changes (calcium carbonate, metal hydroxides)
- adsorption of dissolved trace components on the container walls
- changes in parameters as a result of microbiological activity (e.g. pH-value, oxygen, carbon

dioxide, biochemical oxygen demand, trace organics).

Such processes are normally more rapid in waste and surface waters than in ground or drinking waters.

Some important preservation methods are listed in Table 17.

Table 17: Conservation of substances present in water

| Parameter | Preservation Method | Maximum storage time |
|---|---|---|
| trace metals | 5 mL $HNO_3$/L | several weeks |
| ammonium, total nitrogen | 5 mL $HNO_3$/L | few days |
| mercury | 1 mL 10% $K_2Cr_2O_7$-solution/L | few days |
| nitrite | no stabilization possible, cooling to 4 °C | 1 day |
| cyanides | alkali to pH = 11 | 1 day |
| iron-II | addition of 2,2-bipyridine | 1 day |
| sulfide | 2 mL 10% zinc acetate solution/L | 1 week |
| phenols | alkali to pH = 12 | 1 week |

These methods are only recommendations - the samples must be examined at the earliest possible time. For a whole series of further parameters, especially in relatively pure water, cooling to 4°C is generally a suitable storage condition. Other parameters such as chemical and biochemical oxygen demand can however change rapidly so that only freezing to -18 °C can be considered a suitable method. In this case, only plastic containers are suitable as glass containers would break. It is important that freezing and thawing are rapidly carried out. A pre-cooling in ice water is to be recommended. On freezing, irreversible precipitation can occur thereby falsifying results.

Soil samples should be rapidly dried when possible, as long as no tests are to be performed in the original state.

## 4.6 Sample Collection Procedure

### 4.6.1 Groundwater

Groundwater samples are generally taken from tube wells or shafts with pumps or buckets. The use of pumps is to be preferred. In the case of tube wells, the pump has to be introduced down to the filter area (Figure 37).

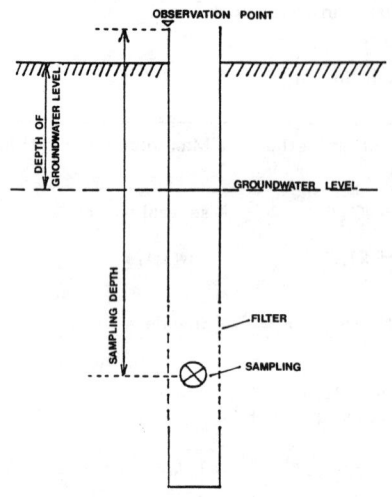

**Fig. 37:** Schematic representation of a tube well

Owing to the fact that very different water characteristics can be encountered within the ground water channel, samples should be taken at various depths. However, the well should not only be equipped with a filter at the base but throughout its length.

Before and after sampling, the water level in the channel should be determined by means of a light lead as conclusions can be reached concerning the permeability of the aquifer.

Before the water samples are bottled, a minimum pumping time should be allowed in order to avoid filling of stationary well water. This time depends on the capacity of the pump and the calibre of the well, half an hour being generally sufficient. The conductivity can also be continually monitored and filling begun after constant conductivity is reached. During pumping, the exit tube is fed to a bucket some meters from the well in order to monitor colour changes.

Should the water be delivered with particles or sand, it must be filtered through coarse filter paper in a large glass or plastic funnel before the physico-chemical tests are carried out. The sample should be filled into several bottles at the place of collection avoiding later dividing in the laboratory. Every bottle must be carefully labelled.

All sampling devices should be dried after use where microbiological testing is involved. Growth of algae, fungi and bacteria is thereby prevented. The exit point must also be cleaned or, where possible, flamed with a gas burner before filling. During collection after a running time of at least 5 minutes a sterile glass container of 100 to 1000 mL capacity is filled in free fall. The sampler should position himself downwind of the container and should avoid coughing or speaking. The open edges of the bottle or the stopper must not be contacted with the fingers. Filling is carried out leaving an air bubble of about 2 mL which allows later shaking to be carried out.

### 4.6.2 Surface Water

The collection of surface water samples normally presents no serious difficulties. The choice of sampling technique depends on the test goals. Environmental contamination or quality control samples are normally taken manually in a random manner. Further examinations of water characteristics normally require larger sample series and automatic samplers can be employed for this purpose. Instead of sample series, depth or area integrated samples can be obtained when single samples collected continuously or discontinuously at various depths or places are combined.

It is normally sufficient for collection that the bucket or sample container is dipped into the water. In flowing waters, the container movement should be against the stream. During collection, stoppers or caps should be placed on a clean surface. On sampling at bridges attention should be paid to the possibility of strong currents near the supports, affecting the water quality (e.g. $O_2$ content). Samples for bacteriological examination should be taken in such a way that the sterile container opening is held against the current. In a still water, the container is pushed through the water and it must be ensured that there is no contact between the hand and the water in front of the container opening.

### 4.6.3 Drinking Water

The collection of drinking water samples for physico-chemical examinations is generally problem-free as taps are usually available. It is recommended that several bottles be filled. For bacteriological tests, the tap must be of metal so that it may be flamed with a gas burner. Still water contained in the whole piping system must be removed before sample collection. A running time of 15 to 30 minutes normally suffices for this purpose.

Sterile glass bottles of 100 to 1000 mL capacity are used and contamination of the bottle necks as well as speaking and coughing during collection are to be avoided. The bottles are filled almost to capacity leaving an air bubble of about 2 mL in volume.

### 4.6.4 Waste Water

The representative collection of crude waste water samples with their variable quantities of suspended materials presents problems, especially where automatic devices are employed. Therefore, an exact measurement of the solids must be dispensed with or pools are prepared from representative random samples. In those cases where organics (e.g. oil) are present in a separate phase manual sampling techniques provide the only suitable collection method.

The collection of purified waste water is, however, relatively easy to carry out and is similar to that for surface water. Random or average samples (time or flow proportional) can be obtained either manually or automatically. In general, 15 minutes-random samples suffice for manual collection. For quality control purposes, two-hourly pooled samples or daily pools are employed. Several bottles are normally filled to avoid later splitting in the laboratory.

### 4.6.5 Soil

Soil samples can be taken either with disturbed or undisturbed stratification. The former were taken without consideration of the natural texture (structure). Undisturbed samples are taken from the earth by means of a cutting cylinder in such a way that their original condition (structure) is maintained.

The soil sample must be representative of the whole area to be examined. This condition is not easy to satisfy even on mixing many single samples since spatial heterogeneity can be much more extreme than in the case of water. For collection, a random distribution of sampling points would be ideal but in relatively homogeneous soil pooled samples of almost equally good quality can be obtained with less effort by reducing the sampling area. Such sampling techniques are illustrated in Figure 38.

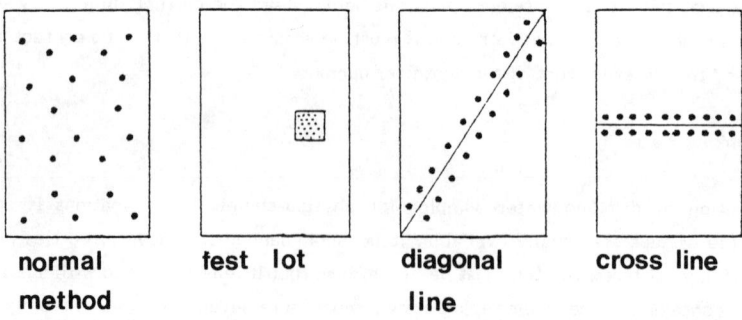

normal method     test lot     diagonal line     cross line

**Fig. 38:** Sampling of soil

For the sampling of soil at least 20 single samples per 10 000 $m^2$ must be taken with an earth boring tool (or spade) and combined to a mixed sample. The usual sampling depth is up to 20 cm in arable land or 10 cm in pasture. Undisturbed soil samples are obtained with a cutting cylinder with minimum capacity of 100 $cm^3$. For special examinations (e.g. testing of nutrient penetration), samples are taken down to 1 m depth after a pit has been dug out to reveal a smooth vertical earth profile.

Samples for nutrient testing should be taken where possible at the same time of year at best after harvest and before application of fertilizer. Sampling is also possible during the main growth period.

For sampling to determine the water profile of soil, a thorough profile description is necessary (e.g. structure, presence of roots, layer changes). The cutting cylinder is pushed steadily in the vertical or horizontal direction in order to avoid compression.

After collection, the samples are packed in plastic bags. Each sample is provided with an identification tag.

For longer storage, disturbed samples are air dried, milled (stones removed) and sieved. The fine earth (particles $<$ 2 mm) is then used in tests.

## 5  Field Measurements

The quality of test results depends to a large extent on the completeness of the information collected in the field, in addition to the sample collecting itself. Furthermore, the measurements to be made locally are of prime importance. These are usually connected with parameters which are likely to change during transport and storage.

### 5.1  Check List

The use of a check-list during the local survey and sample collection should not be dispensed with even under difficult field conditions as all collected information may be important for the subsequent assessment. For a number of reasons, even experienced sample collection personnel can make mistakes or forget important steps. The following check-list is not an exhaustive list of points but should function only as a memory aid for the collection of information in the field. The facts thus obtained are entered into the sample collection report (Table 18).

### 5.2  Parameters

#### 5.2.1  Sensory Examination

The sensory examination should take place during sample collection since changes can occur during tansport and storage. The tests include odor, taste, clarity, turbidity and color. Soil samples are tested for odor, color and general "feeling" during hand contact.

The odor test should be performed immediately after sampling as certain odors such as that of hydrogen sulfide can disappear rapidly. Odor strengths and types may be designated as follows:

| Odor strength: | Odor type: |
|---|---|
| very weak, weak | earthy, mossy, turfy |
| clear, strong, very strong | musty-foul, dungy, fishy, aromatic. |

The color of water may be tested with colorimeter tubes for comparison purposes. The visual designation of color is as follows: colorless, very weakly colored, weakly colored, strongly colored.

In addition, the corresponding tone is given, e.g.: yellowish, yellowish-brown, brownish, yellowish-green etc.

## 5.2 Parameters

**Table 18:** Check-List for sampling and observation

| Parameter | Ground Water | Surface Water | Drinking Water | Waste Water | Soil |
|---|---|---|---|---|---|
| Position (M,O) | x | x | x | x | x |
| Entry of Coordinates (C) | x | x | x | x | x |
| Geological conditions (C, O) | x | x | x |  | x |
| Description of area (O) | x | x | x | x |  |
| Structure of surface (O) | x | x |  |  | x |
| Use of land (O) | x | x | x | x | x |
| Undergrowth (O) | x | x |  |  | x |
| Flow velocity (O, M) | (x) | x | (x) | x |  |
| Flow (O, M) |  | x | x | x |  |
| Sedimentation (O, M) |  | x |  | x |  |
| Description of water system (O) e.g. entrances |  | x |  | x |  |
| - organisms | x | x | x | x |  |
| - eutrophy |  | x |  |  |  |
| - visible contamination | x | x | x | x |  |
| - type of spring or well | x | x |  |  |  |
| - corrosion signs | x | x | x | x |  |
| - gas development | x | x |  | x |  |
| Description of soil (O) e.g. color |  |  |  |  | x |
| - type |  |  |  |  | x |
| - concretions |  |  |  |  | x |
| - density |  |  |  |  | x |
| - roots |  |  |  |  | x |
| - humidity |  |  |  |  | x |
| Measurements (M) |  |  |  |  |  |
| - air temperature | x | x | x | x |  |
| - air pressure | x | x |  | x |  |
| - color, odor | x | x | x | x | x |
| - taste | (x) |  | (x) |  |  |
| - turbidity | x | x | x | x |  |
| - sight depth |  | x |  |  |  |
| - sedimentation | x | x | x | x |  |
| - precipitation | x | x | x | x |  |
| - pH-value | x | x | x | x |  |
| -redox potential | x | x | x | x |  |
| - electrical conductivity | x | x | x | x |  |
| - oxygen | x | x | x | x |  |
| - chlorine |  |  | x |  |  |
| - carbon dioxide | x | x | x | (x) |  |
| - aggressiveness | x | (x) | x | (x) |  |

M = Measurement
O = Observation
C = Chart

The examination of taste is only carried out when no suspicion of possible infectious contamination exists. Taste sensations may be designated as follows: tasteless, salty, bitter, alkaline, sour, astringent, metallic, repulsive.

The apparent degrees of tase may be differentiated by "weak", "clear" and "strong".

The visible depth of water is determined as that depth at which a white disc let down with a line or pole is just visible. Down to 1 m, the values are given in cm intervals, and beneath 1 m, 0.1 m intervals are read off.

The simple test of turbidity is carried out by filling a clear glass container (1 Liter) about two-thirds full, shaking well and comparing against a black and then a white background. The following degrees of turbidity are determined: clear, opalescent, weakly turbid, strongly turbid, opaque.

A more exact definition is obtained by comparing the turbidity against a series of standard silica suspensions (1 and 0.1 g/L $SiO_2$) either visually or with the aid of a photometer.

### 5.2.2 Temperature

Areas of Application ⟶ Water, waste water, soil

Apparatus
a) air temperature
calibrated mercury thermometer with 0.5 °C graduations, measurement range - 20 to + 60 °C.
b) water and soil temperature
calibrated mercury thermometer with 0.1 °C graduations, measurement range 0 to 100 °C.

For waste water measurements a maximum thermometer may be used.

Measurement
The air temperature measurement is carried out with a dry thermometer, approximately 1 m above the sampling point. The thermometer must be shaded from the sun.

The water temperature is measured by dipping the thermometer to the reading off depth and waiting until the reading is constant. Where the direct measurement is not possible (e.g. in springs), a larger quantity of water is collected and the temperature taken as soon as possible.

The measurement of soil temperature should be carried out at the relevant depth as significant temperature gradients usually exist.

## 5.2.3 Settleable Matter

Water and waste waters contain sedimentable components. The determination should be carried out soon after sampling in order to avoid errors through flocculation. The method is suitable for the measurement of sedimentable components above 0.1 ml/L.

Areas of Application → water, waste water

Apparatus
Imhoff sedimentation glass, 1 L
Holding device

Measurement
1 L of the shaken sample is placed into the sedimentation glass immediately after collection. After approxemately 50 minutes and 110 minutes the container is rotated about the vertical axis to cause any materials clinging to the walls to sink to the bottom. After 1 to 2 hours the volume of sediment is read off.

Calculation of Results
The recorded values are rounded off as follows:

**Table 19:** Calculation of sediments from recorded values

| Recorded value (mL) | Rounded value (mL/L) |
|---|---|
| <2 | to 0.1 |
| 2 to 20 | to 0.5 |
| 10 to 40 | to 1 |
| >40 | to 2 |

## 5.2.4 pH-Value

The pH value is the negative normal logarithm of the hydrogen ion activity (mol/L) and has a value of 7 at 25 °C in pure water (neutral point). As a result of the presence of acids and alkali and also hydrolysis of dissolved salts, the pH value can decrease (acid) or increase (alkali). The presence of salts of strong bases and weak acids (e.g. $Na_2CO_3$) increases the pH values; salts of weak bases and strong acids (e.g. $CaCl_2$) cause decreases. In soils, acidification is a result of hydrolysis of iron and aluminium compounds as well as humic acid formation from the degradation of organic materials.

The pH values of neutral waters usually lie between 6.5 and 7.5 and lower values are a result of free $CO_2$. Biogenic decalcification in surface waters can cause the pH value to reach 9.5.

Areas of Application ⟶ Water, waste water, soil

Apparatus
a) Indicator papers
b) pH-meter with single-rod glass electrode

Reagents and Solutions
Buffer solutions pH 4.62, 7.0, 9.0

pH 4.62:   200 mL of 1 M $CH_3COOH$ is mixed with 100 mL 1 M NaOH and 700 mL $H_2O$

pH 7.0:   a) 9.078 g $KH_2PO_4$ with $H_2O$ to 1 L
          b) 11.88 g $Na_2HPO_4 \cdot 2\ H_2O$ with $H_2O$ to 1 L
          2 parts of a) are mixed with 3 parts of b).

pH 9.0:   a) 12.40 g $H_3BO_3$ and 100 mL 1 M NaOH to 1 L with $H_2O$
          b) 0.1 M HCl
          8.5 parts of a) are mixed with 1.5 parts of b)

Sample Preparation
Water samples require no preparation for pH measurements if the electrical conductivity is below 100 $\mu S/cm$. Various methods are employed for soil. The pH value can be measured in a soil paste or in a suspension. Suspensions are prepared by shaking 1 part of soil with 2.5 parts of either distilled water or a 0.1 M KCl solution for 1 hour and the pH is then measured. Measurements with 1 M KCl generally deliver values which are 0.5 to 1 pH unit lower than those in distilled water.

Calibration and Measurement
pH measurements with indicator papers are carried out by dipping the paper strip into the solution and comparing the resulting color with the manufacturers standards after the given development period. The papers in use today generally have dyes which do not fade during the measurement time.

Before performing potentiometric pH measurements, new or dry manufacturers glass-electrodes must be placed in water or 3 M KCl for several days before use (see manufacturers instructions). The calibration is carried out with two standard buffer solutions. The pH of the sample should lie between these values. The sample temperature is determined at the same time and is entered into the meter to allow for a temperature correction. Where samples having very different pH values are measured successively the electrode should be treated with water for some time between measurements. The reading is taken after the indicated value remains constant for about one minute. The value can be read to within 0.1 units; sensitive instruments are exact to 0.01 units.

### Interfering Factors

At pH values above 10, so-called alkali errors may occur. In such cases, the use of an alkali resistant electrode is recommended. Changes in glass structure can occur in older electrodes so that errors may appear when measuring in weakly buffered waters. In such cases, the electrode should be renewed.

The sensitivity can be reduced by the presence of oil in the samples. Measurement errors in oil-containing waters may be prevented by washing the electrode before each measurement using soap or detergents followed by water, dilute hydrochloric acid and finally with more water.

### Calculation of Results

Below pH 2 and above pH 12, the result is given to the nearest 0.1 unit. Otherwise, depending on the type of instrument, the result can be accurate to the second decimal place.

## 5.2.5 Redox Potential

The redox-potential is an electrode-dependent parameter giving a measure of the electron activity in waters. In a similar way to the pH value, redox potentials largely control the chemical processes occurring in nature. Sudden decreases in the redox potential of water and waste water can be brought about by anaerobic biological processes. In still waters, oxygen can enter the surface by diffusion, thereby supporting an aerobic milieu. At the same time, an anaerobic environment can be formed in the depths or in sediments through the lacking of intermixing. For this reason the redox potential often delivers the first evidence of changes in the water properties. The redox potential measures the competing processes of electron donation (reduction) or acceptance (oxidation).

### Areas of Application ⟶ water, waste water

### Apparatus
pH/mV-meter
measuring rod with Pt or Au electrodes

### Reagents and Solutions

Redox buffer solution: Chinhydron is dissolved in a pH buffer to saturation. The solution must always be freshly prepared. The redox potential is always proportional to the pH value in the range 1 to 7

e.g.: pH 4.62 ⟶ 427 mV
pH 7.00 ⟶ 285 mV (at 25 °C)

### Calibration and Measurement

Control measurements with the aid of the chinhydron-buffer solution should be carried out in regular intervals. For measurement, the sample is placed in a container and the electrode set dipped into it. The reading is recorded when the value shown does not change over a period of several minutes. Changing between solutions with very different ionic activities leads to very long delays in determining the end point.

Interfering Factors

The recorded value depends on many factors such as the ionic activity, the temperature and the nature of the electrode surface. When slow reading is encountered, a careful cleaning of the metal ring with talcum often helps.

Calculation of Results

mV values should only be used to give some idea of the presence of aerobic or anaerobic processes. Below -200 mV processes are strictly anaerobic, between 0 and -200 mV they show transitional character and positive mV-values indicate aerobic processes. A direct comparison of values can only be valid for the same redox pairs, ionic strengths and pH-values. The pH-value has always to be given.

### 5.2.6 Electrical Conductivity

The electrical conductivity is a total parameter for dissolved, dissociated substances. Its value depends on the concentrations and degrees of dissociation of the ions as well as the temperature and migration velocity of the ions in the electric field.

No conclusions can be drawn about the types of ions present. However, exact conclusions can be drawn about the concentrations of dissolved electrolytes from the electrical conductivity when the ionic composition and equivalent conductivities are known.

Electrical conductivity measurements are often employed to monitor desalination plants and are further used to test surface and ground waters. In coastal regions, conductivity data can be used to decide the extent of penetration of sea water into the ground water. In soil examinations, conductivity provides information on the proportion of soluble salts and therefore on the usefulness of the soil.

The conductivity is represented by the reciprocal value of the electrical resistance in Ohms relative to a cubic centimetre of water at 25 °C (unit: $\mu S/cm$ with $1\ S = \Omega^{-1}$).

Areas of application ⟶ water, waste water, soil

Apparatus

conductimeter with measuring cell
thermometer

Reagents and Solutions

KCl solution 1: 7.456 g anhydrous KCl is made up to 1 L with water

KCl solution 2: 100 mL of solution 1 is made up to 1 L with water

Sample Preparation

No sample preparation is necessary with water samples. Extracts (1 : 5) are normally prepared from soil before measuring the conductivity. For this purpose the soil sample is shaken repeatedly with the corresponding amount of demineralised water in a closed container. Between shaking steps the sample is left for 30 minute periods. Afterwards the sample is filtered.

Calibration and Measurement

Before measurement, the container and cell must be washed several times with the solution under test. The measurement should be carried out at 25 °C. Otherwise, the effect of temperature must be compensated according to the manufacturers instructions. When the sample and apparatus have the same temperature, the temperature is read off from the thermometer and then the conductivity from the meter.

The conductimeter must be tested from time to time with the aid of the KCl solutions to measure the cell constants. This procedure should also be followed according to the instructions delivered with the instrument.

Calculation of Results

The ionic concentration of a solution influences the conductivity as well as the ionic strength. The following relationship allows the ionic strength I to be calculated from the electrical conductivity

$$I = 1.83 \; \chi_{20} \cdot 10^{-5}$$

$\chi_{20}$ electrical conductivity in $\mu S/cm$ at 20 °C

This formula is only valid for waters containing carbonate.

In the case of soil, the calculation is only approximate owing to the variety of salts present.

As an approximation:

$$I = 10^{-3} \; S \; cm^{-1} \text{ is about 64 mg/L salt/100 mL extract.}$$

## 5.2.7 Oxygen

The presence of oxygen is essential for the survival of most organisms in water. This is also true for the metabolic pathways of aerobic bacteria and other microorganisms which are responsible for the degradation of pollutants in water and for this purpose utilize oxygen as an electron acceptor.

Oxygen reaches the water via surface diffusion as well as in photosynthetic processes occurring in algae and submerged plants. Where the plant production is great, oxygen oversaturation can occur. In drinking water, a minimum amount of oxygen ( >4 mg/L) is essential to prevent corrosion of the carrier pipes.

The oxygen may be determined amperometrically or titrimetrically according to the modified Winkler method.

Areas of Application ⟶ Water, waste water

Apparatus
a) amperometric determination:
   oxygen measuring apparatus

b) Winkler method (titrimetric):
   ground glass bottles of known capacity (110 to 150 mL)
   glass apparatus for volumetric analysis

Reagents and Solutions
a) amperometric method:

| | |
|---|---|
| zero solution: | freshly prepared saturated sodium dithionite ($Na_2S_2O_4$) solution |
| air saturated solution: | distilled water is aerated |

b) Winkler method

| | |
|---|---|
| manganese-II-sulfate solution: | 480 g $MnSO_4 \cdot 4 H_2O$ (or 400 g $MnSO_4 \cdot 2 H_2O$) made up to 1 L with water |
| alkaline iodide-azide solution: | 350 g NaOH (or 500 g KOH) and 150 g K I (or 135 g Na I) and 1 g $NaN_3$ are made up to 1 L with water |
| 0.01 M sodium thiosulfate solution: | freshly prepared by dilution of 0.1 M sodium thiosulfate |
| orthophosphoric acid: | at least 85 wt.% |
| starch solution: | 1 g soluble starch is boiled and a few drops of formalin are added |

Sample Preparation
For an amperometric measurement, the electrode can be placed directly into the solution. In the Winkler method, the test bottle is filled bubble-free with a tube and washed through several times by overflow.

## 5.2 Parameters

Calibration and Measurement

a) amperometric

The oxygen electrode is placed first of all in the control solution until a constant reading is obtained. The electrode is then washed and placed into the air saturated solution and the constant reading is recorded. Further adjustment is based upon the air pressure and water temperature according to Table 20:

**Table 20:** Oxygen saturation β ($O_2$) of water depending on water temperatures and air pressure

| Water temperatures (°C) | β ($O_2$)-concentration (mg/L) at air pressure (hPa ≙ mbar) |       |       |       |       |
|---|---|---|---|---|---|
|    | 933   | 960   | 986   | 1013  | 1040  |
| 0  | 13.41 | 13.80 | 14.18 | 14.57 | 14.95 |
| 2  | 12.70 | 13.06 | 13.43 | 13.79 | 14.16 |
| 4  | 12.04 | 12.38 | 12.73 | 13.08 | 13.42 |
| 6  | 11.43 | 11.76 | 12.09 | 12.42 | 12.75 |
| 8  | 10.87 | 11.19 | 11.50 | 11.81 | 12.13 |
| 10 | 10.36 | 10.66 | 10.96 | 11.26 | 11.56 |
| 12 | 9.88  | 10.17 | 10.46 | 10.74 | 11.03 |
| 14 | 9.45  | 9.72  | 9.99  | 10.27 | 10.54 |
| 16 | 9.04  | 9.31  | 9.57  | 9.83  | 10.10 |
| 18 | 8.67  | 8.92  | 9.18  | 9.43  | 9.68  |
| 20 | 8.33  | 8.57  | 8.81  | 9.06  | 9.30  |
| 22 | 8.01  | 8.24  | 8.48  | 8.71  | 8.95  |
| 24 | 7.71  | 7.94  | 8.16  | 8.39  | 8.62  |
| 26 | 7.43  | 7.65  | 7.87  | 8.09  | 8.31  |
| 28 | 7.17  | 7.38  | 7.60  | 7.81  | 8.02  |
| 30 | 6.93  | 7.13  | 7.34  | 7.55  | 7.76  |
| 32 | 6.70  | 6.90  | 7.10  | 7.30  | 7.50  |
| 34 | 6.48  | 6.67  | 6.87  | 7.07  | 7.26  |
| 36 | 6.27  | 6.46  | 6.65  | 6.84  | 7.03  |

(conversion factor from mbar to Torr: 0.750)

The electrode is then ready for measurements.

b) Winkler method

For oxygen-fixation the following are pipetted successively under the sample surface in a full bottle: 0.1 mL manganese sulfate solution, 0.5 mL K I solution. The container is closed free of air bubbles and shaken. In the laboratory, 2 mL of phosphoric acid are also pipetted, the bottle closed and the shaking continued. After approx. 10 minutes the content is transferred to an Erlenmeyer flask and titrated with 0.01 M thiosulfate solution. When only a weakly yellow color remains, 1 mL of the starch solution is added and titration continued until the blue color disappears.

### Interfering Factors

When the control values or saturated value are found to lie outside the normal measuring range, the inner electrode, working electrode and membrane must be cleaned or replaced. The main interfering substances is hydrogen sulfide. Interferences in the Winkler method through iron III and nitrite ions may be prevented by addition of phosphoric acid or azide during the determination. The iodine difference method is recommended in those cases where larger amounts of oxidizing or reducing agents are present (see literature list: Deutsche Einheits-verfahren; US-Standard Methods).

### Calculation of Results

1 mL thiosulfate-solution corresponds to 0.08 mg oxygen. Content of dissolved oxygen:

$$O_2 \text{ (mg/L)} = \frac{x \cdot 80}{V - V_R}$$

where
- $x$ = consumption of 0.01 M sodium thiosulfate solution (mL)
- $V$ = volume of sample container (mL)
- $V_R$ = volume of reactives added (without phosphoric acid)

Oxygen saturation:

$$O_2 \text{ (\%)} = \frac{a \cdot 100}{b} \text{ \%}$$

where
- $a$ = measured oxygen concentration (mg/L)
- $b$ = oxygen saturation concentration at measured temperature (mg/L)

### 5.2.8 Chlorine

Chlorine is used as a disinfectant to treat drinking water, bathing waters and sometimes also waste water. Chlorine in the form of dissolved elementary Cl, chloric acid (HOCl) or hypochlorite ions is referred to as "free chlorine" whereas chlorine in compounds formed by reaction of hypochlorite ions with ammonium or organic compounds containing amine groups is known as "bound chlorine". Both together form "active chlorine". The differentiation between the forms is advantageous since free chlorine is the stronger disinfectant and thus delivers better information for the water treatment. The reaction of chlorine with substances carried in water is shown schematically in Figure 39.

Active chlorine should be determined at each stage in the processing of drinking water and in the mains in order to guarantee a bacteriologically impeccable water. Active chlorine should be present in drinking water within the range 0.2 to 0.5 mg/L.

## 5.2 Parameters

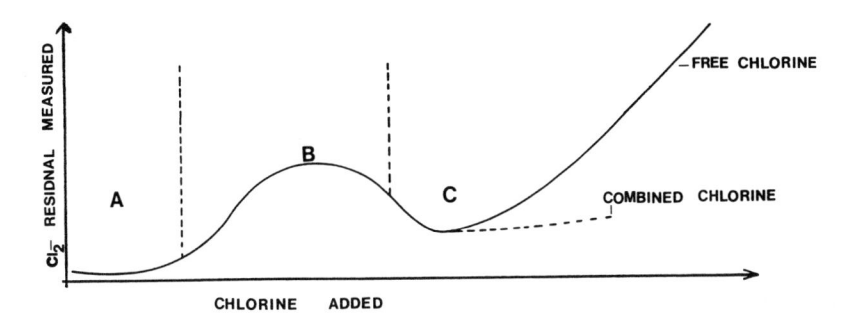

**Fig. 39:** Stages of reaction between chlorine and substances present in water
- A: consumption of chlorine
- B: formation of organic chlorine compounds
- C: degradation of organic chlorine compounds

The following description includes a simple comparator field method and the titrimetric DPD method (diethyl-p-phenylendiamine, $C_{10}H_{16}N_2$).

Areas of Application ⟶ water, waste water

### Apparatus

a) Field method

Comparator (e.g. Lovibond-Tintometer$^R$)
Comparator filters B ($Cl_2$): 0.1 to 1 and 1 to 4 mg/L $Cl_2$
DPD tablets No 1 and 3 (producer: Lovibond-Tintometer, UK)

b) Titrimetric method

Glass equipment for wet analysis

### Reagents and Solutions

Glycine solution: 20 g glycine ($C_2H_5NO_2$) are dissolved in 200 mL water.

Buffer solution: 24 g disodium hydrogen phosphate ($Na_2HPO_4$) and 46 g potassium hydrogen phosphate ($KH_2PO_4$) are dissolved in approx. 800 mL water. 100 mL of a 0.8% EDTA solution ($C_{10}H_{14}N_2O_8Na_2 \cdot 2\,H_2O$) are added and the solution made up to 1 L. The solution is stored in brown bottles and may no longer be used when colored.

DPD solution: 1.5 g DPD sulfate is dissolved in 800 mL water, 8 mL 40% sulfuric acid and 25 mL 0.8% EDTA-solution are added and filled up to 1 L. The solution is stored in brown bottles and may no longer be used when colored.

FAS solution: 1.106 g ferrous ammonium sulfate solution $((NH_4)_2Fe(SO_4)_2 \cdot 6 H_2O)$ is dissolved in 800 mL water. 1 mL 40% sulfuric acid is added and made up to 1 L. The solution can be used for approx. 1 month. Titration is carried out with potassium dichromate solution (see section 6.2.1.5)

Potassium iodide: solid

## Calibration and Measurement

### a) Field method with the Lovibond Tintometer[R].

The measurement cuvette is washed with the sample water and the No 1 tablet dissolved in a little water. This solution is then made up to 10 mL, mixed and placed together with the comparison cuvette containing pure water into the comparator. The apparatus is held against the light and adjusted by means of the thumb-wheel until the colors are identical. The reading is noted.

Then the No 3 tablet is placed into the sample cuvette, mixed and left for 2 minutes. The new value giving the same color is noted. The first value gives the free chlorine and the difference between the values gives the bound chlorine.

### b) Titration Method

Titration 1 - Determination of total chlorine:
5 ml glycine solution and 200 mL of water sample are placed in an Erlenmeyer flask and left for 2 minutes. The solution is then poured into a second Erlenmeyer containing 10 mL of buffer solution and 5 mL DPD solution. The contents are then titrated with FAS solution until colorless.

Titration 2 - Determination of free chlorine:
10 mL buffer solution and 5 mL DPD solution are placed in an Erlenmeyer together with 200 mL water sample. After 5 minutes, titration is carried out with FAS solution until colorless.

Titration 3 - Determination of bound chlorine:
The solution after titration 2 is treated with 1 g solid potassium iodide. After dissolving and leaving for 5 minutes, the solution is titrated with FAS solution until colorless.

## Interfering Factors

Large amounts of manganese oxide (higher oxidation state) affect the results obtained. Higher concentrations of copper and iron ions which would normally also interfere are bound by the added EDTA solution.

## Calculation of Results

free chlorine (mg/L) = (consumption from Titration 2 / consumption from Titration 1) · 0.5 mg/L bound chlorine (mg/L) = (consumption from titration 3) · 0.5 mg/L

Where chlorine dioxide is employed as the chlorinating agent in drinking water, the following equation applies:

chlorine dioxide (mg/L) = (consumption from titration 1) · 0.95 mg/L

## 5.2.9 Alkalinity (Acidic Capacity)

The acidic capacity ($K_A$ in mmol/L) in is defined as the capacity of substances contained in the water to take up hydroxonium ions ($H_3O^+$) to reach a defined pH value. In natural waters, the hydroxonium ions are bound mainly by the anions of weak acids (mainly carbonate and hydrogen carbonate). In addition, the suspended calcium carbonate in chalk-rich waters can be determined at high pH values by the acidic capacity.

Two values should be distinguished:
  acidic capacity to pH 8.2 ⟶ $K_{A\,8.2}$
  acidic capacity to pH 4.3 ⟶ $K_{A\,4.3}$

$K_{A8.2}$ is determined in waters having a pH value above 8.2 and $K_{A4.3}$ in those waters with pH values above 4.3. Depending on the accuracy requirements, the measurement can be made potentiometrically or with the aid of color indicators.

Areas of Application ⟶ water, waste water

Apparatus
pH meter
electrode set
magnetic stirrer
100 mL measuring cylinders or pipettes
burettes
300 mL Erlenmeyer flasks.

Reagents and Solutions
0.1 M HCl
0.02 M HCl

| | |
|---|---|
| Phenolphthalein indicator solution: | 1 g phenolphthalein is dissolved in 1000 mL of a water-ethanol (1 : 1) mixture |
| Mixed indicator: | 0.02 g methyl red and 0.1 g bromocresol green are dissolved in 100 mL ethanol |

## Sample Preparation

The end-point determination with color indicators can be hindered by the presence of color and turbidity. Free chlorine can also destroy the indicator. Contaminants causing turbidity can be removed by filtration (filter 0.45 µm) and coloring can normally be significantly reduced by addition of activated charcoal followed by filtration. Free chlorine is reduced by adding a drop of 0.1 M sodium thiosulfate solution.

## Measurement

100 mL sample is placed into an Erlenmeyer flask. With potentiometric measurement, the pH electrode is placed into the water and titration carried out with 0.1 M HCl until the pH has reached 8.2 and the mixture is then left for 2 minutes. The volume of acid solution consumed is noted and titration continued to pH 4.3. During titration the solution is stirred or shaken. Where the acid consumption is very small, the titration is repeated with 0.02 M HCl.

For measurements involving a color indicator, 2 or 3 drops of phenolphthalein solution are added. On obtaining a red color, the sample is first of all titrated until colorless, 2 to 3 drops of the mixed indicator are then added and the titration continued until the end point is reached (green to red).

## Interfering Factors

Interference caused by color, turbidity or free chlorine have already been discussed above. In addition, absorption or loss of carbon dioxide during or after sample collection can falsify the results. Dissolved silicates, phosphates, borates or humic acid salts are also accounted for in the result. Their presence does not constitute an interference but must be considered when calculating the concentration of carbon dioxide, hydroxide or carbonate.

## Calculation of Results

$K_{A\ 8.2}$ and $K_{A\ 4.3}$ are calculated as follows:

$$K_{A\ 8.2} = \frac{V_1 \cdot c \cdot 1000}{s} \quad (mol/L)$$

$$K_{A\ 4.3} = \frac{V_2 \cdot c \cdot 1000}{s} \quad (mol/L)$$

where $V_1$ and $V_2$ are the volumes of acid consumed in mL in order to reach pH 8.2 and 4.3, respectively, c is the acid concentration in mol/L and s is the volume of the sample in mL.

The numerical value for acid consumption using 0.1 m HCl and 100 mL sample volume gives the acidic capacity $K_{A\ 8.2}$ or $K_{A\ 4.3}$ directly.

### 5.2.10 Acidity (Base Capacity)

The base capacity ($K_B$, in mmol/L) is defined as the capacity of substances contained in the water to take up hydroxyl ions (OH⁻) to reach a defined pH value. In natural waters, this occurs mainly through dissolved carbon dioxide so that determination of base capacity can be used under certain conditions to determine the content of this component.

Two values are distingiushed:

base capacity to pH 4.3 → $K_{B\ 4.3}$

base capacity to pH 8.2 → $K_{B\ 8.2}$

$K_{B\ 8.2}$ is mainly determined although in special cases (peat water, open drain water, water from cation exchangers) $K_{B4.3}$ is determined. Depending on the accuracy requirements, the measurements can be carried out potentiometrically or with the aid of color indicators.

Areas of Application ⟶ water, waste water

Apparatus
pH meter
set of electrodes
magnetic stirrer
100 mL measuring cylinder or 100 mL pipette
burettes
300 mL Erlenmeyer flasks

Reagents and Solutions
0.1 M NaOH
0.02 M NaOH

| | |
|---|---|
| Phenolphthalein indicator solution: | 1 g phenolphthalein in 100 mL of an ethanol-water (1 : 1) mixture |
| Methyl orange solution: | 0.05 g methyl orange is dissolved in 100 mL water |
| Potassium-sodium tartrate solution: | 50 g K-Na tartrate ($C_4H_4O_6KNa \cdot 4\ H_2O$) is made up to 100 mL with water. After 1-2 days the pH is adjusted to 8.2 |

Sample Preparation
Color, turbidity and also free chlorine can interfere with the determination. The sample preparation can be carried out as described under section 5.2.9 "Acidic Capacity". Losses of carbon dioxide during sample preparation should be avoided (e.g. by submerged pipettes).

Measurement
100 mL of the water sample is carefully filled into an Erlenmeyer flask -the sample is allowed to run slowly along the wall from the measuring cylinder or pipette. Potentiometric titrations are carried out with 0.1 M NaOH to pH 4.3 (where sample pH is less than 4.3) with careful mixing or swirling. In the case where the pH of the water lies above 4.3, approximately 1 mL

K-Na tartrate solution is added and the mixture then titrated to pH 8.2. The final pH value should remain constant for at least 2 minutes. The volume of acid used is noted and the procedure repeated with 0.1 M or 0.02 M NaOH, depending on the level of consumption, whereby the total amount of solution used in the first determination is added at once.

The determination with colored indicators is carried out by titrating $K_{B\ 4.3}$ after adding 2 to 3 drops of methyl orange (indicator change from orange to yellow) and separately $K_{B\ 8.2}$ after adding 5 drops of phenolphthalein solution (indicator change from colorless to pink).

Samples containing high concentrations of carbon dioxide are placed directly into graduated titration containers (or graduated beakers or volumetric cylinders). A defined volume of 0.1 M NaOH is added together with 1 mL K-Na tartrate solution and titrated back with 0.1 M HCl.

Interfering Factors

Interference caused by the presence of oxidizable or hydrolyzable ions of iron, aluminium or manganese is prevented by the addition of Na-K tartrate. Loss of carbon dioxide during sample collection and titration can reduce the value of $K_{B\ 8.2}$.

Calculation of Results

$K_{B\ 4.3}$ and $K_{B\ 8.2}$ are calculated as follows:

$$K_{B\ 4.3} = \frac{V_1 \cdot c \cdot 1000}{s} \quad (mol/L)$$

$$K_{B\ 8.2} = \frac{V_2 \cdot c \cdot 1000}{s} \quad (mol/L)$$

where $V_1$ and $V_2$ are the volumes of NaOH consumed in order to reach pH 4.3 and 8.2, respectively, c is the concentration of NaOH in mol/L and s is the sample volume in mL.

With a sample volume of 100 mL and use of 0.1 M NaOH, the numerical value of alkali consumed gives a direct measure of the base capacity $K_{B\ 4.3}$ and $K_{B\ 8.2}$.

## 5.2.11 Calcium Carbonate Aggression

Prior to the treatment of crude water in the preparation of drinking water, but also during the various processing stages it is important to know if the water is calcium aggressive or precipitating. The calcium carbonate - carbon dioxide equilibrium between the dissolved carbon dioxide and the solid calcium carbonate may be described by the following equations:

$$H_2CO_3 + H_2O \rightleftharpoons H_3O^+ + HCO_3^-$$

$$HCO_3^- + H_2O \rightleftharpoons H_3O^+ + CO_3^{2-}$$

$$CaCO_3 \rightleftharpoons Ca^{2+} + CO_3^{2-}$$

These reactions cause the solution to have a particular pH value.

Two effects are possible when calcium carbonate solids are added to water which is not in equilibrium:

   the pH value increases i.e. the water is chalk aggressive
   the pH value decreases, i.e. the water is chalk precipitating.

The quick test described below is usually sufficient for the rapid assessment of waters. More exact measurements, e.g. for the dimensioning of deacidification plants, require other methods (Literature list: Frevert; Deutsche Einheitsverfahren; US-Standard-Methods).

Area of Application        water

Apparatus
pH meter
electrode set
conical containers, approx. 50 mL to 100 mL capacity, or even 50 mL beakers
thermometer
sample collection tubing

Reagents and Solutions
Marble powder ($CaCO_3$, 99 %)
dilute hydrochloric acid (ca. 10 %)

Measurement
The pH electrode is dipped into the tip of the conical container. Sample water is passed through until a constant reading is obtained and then marble powder is added to completely cover the electrode ball. After about 2 minutes, the pH is read again. The temperature of the marble should be the same as that of the water and this should be monitored using a thermometer during the measurement. The container is cleaned with dilute hydrochloric acid after each measurement.

Calculation of Results
The difference between the measured pH values is considered where $\pm$ 0.04 pH units indicate a calcium-carbon dioxide equilibrium. (Table 21).

**Table 21:** Measurement of calcium carbonate aggression in water

| pH increase | aggressive | pH decrease | precipitating |
|---|---|---|---|
| + 0.04 to + 0.1 | essentialy equilibrium | -0.04 to -0.1 | essentially equilibrium |
| +0.1 to 0.5 | weakly aggressive | -0.1 to -0.5 | calcium precipitation possible |
| +0.5 to + 1 | aggressive | -0.5 to -1 | calcium precipitating |
| more than +1 | strongly aggressive | more than -1 | strongly calcium precipitating |

# 6 Laboratory Measurements

## 6.1 Sample Preparation

### 6.1.1 Water and Waste Water Samples

Water and waste water samples generally require no special sample preparation. However, particular attention should be paid during collection and the subsequent analytical steps when components which are easily changed are present in the water. Such problems are discussed in detail in section 4. It is generally desirable to carry out the analysis promptly after sample collection in order to minimize falsification of the results.

In some samples, metals such as copper or zinc can be complexed to humus components to various degrees. It is recommended that after addition of nitric acid or a nitric acid-hydrochloric acid mixture (1:3), the sample should be evaporated to a smaller volume in order to break up such complexes. In addition, evaporation to a smaller volume allows the limit of determination to be lowered as long as matrix problems related to increased salt concentration do not arise. Also, metals adsorbed on to suspended particles are usually oxidized and brought into solution by this type of pre-treatment.

If the oxygen demand of suspended solids in waste water is to be included in the measurement of COD or $BOD_5$, the sample must be homogenized (e.g. by means of a high speed stirrer).

Most of the pretreatment steps for water and waste water are described in the specified methods.

### 6.1.2 Soil Samples

Soil samples are normally dried promptly after collection as long as no particular tests involving the water content are to be carried out. The drying process must be carried out mildly in order to prevent the occurrence of rearrangement processes. Larger pieces of earth are crushed enabling roots and other living material to be removed. Most soil tests are performed on the fine fraction which is obtained by sieving through a 2 mm-mesh. Clay-rich soils are sieved before complete drying. For certain determinations e.g. total metals, the sample must be very finely milled.

The more important steps in soil sample preparation are listed below:

<u>Digestion for determination of total Ca, Mg, K, Na, Fe, P</u>

0.5 g air-dried, milled fine soil are heated for a short time at red heat in a platinum crucible. After cooling, the sample is moistened with water and treated with 1 mL perchloric acid (60%) and 10 mL hydrofluoric acid (40%). The contents are reheated to dryness on a sandbath to approx. 180 °C. After cooling, 15 mL hydrochloric acid (10%) are added and the mixture heated in a closed crucible to dryness. If the solution after addition of HCl is not clear, the HF/HClO$_4$ step must be repeated. In this case the solution is transferred to a 100 mL volumetric flask.

### C - Determination after wet ashing (Determination of Chemical Oxygen Demand see 6.2.1.5)

2 g air-dried fine soil (0.5 g if peaty soil) is placed in a 250 mL volumetric flask together with 40 mL conc. sulfuric acid. After standing for 10 minutes at slightly reduced temperatures, the mixture is treated with 25 mL potassium dichromate solution (98.07 g $K_2Cr_2O_7$/L). The flask is placed in a laboratory oven for 3 hours with occasional agitation. After cooling, the flask is topped up to 250 mL and a 25 mL aliquot back-titrated to give the unconsumed potassium dichromate as described in the section "Determination of the chemical oxygen demand".

The end-point determination is improved by adding 5 mL of a special acid mixture before the titration (150 mL conc. $H_2SO_4$ + 150 mL conc. $H_3PO_4$ + 5 g $FeCl_3 \cdot 6\,H_2O$ are mixed and added to some water under cooling. After cooling, the solution is made up to 1 L).

### Nitrogen Content (Determination of Kjeldahl Nitrogen; see 6.2.1.11)

1 to 5 g of air-dried fine soil together with a spatula tip of selenium reaction mixture are placed into a Kjeldahl flask and treated with 6 mL conc. sulfuric acid. The sample is then heated until the residue becomes colorless. After cooling, the solution is transferred to a 1 L round-bottom flask and 25 mL sodium hydroxide solution (30%) is added. The distillation and further steps are then carried out as described under "determination of Kjeldahl Nitrogen".

### Weatherable Ca, K, P

10 g of air-dried fine soil are heated in a porcelain crucible for 1 hour at 500 °C. After cooling, 50 mL hydrochloric acid (30%) are added and the mixture carefully heated on a sandbath, during which the top of the crucible is covered with a watch-glass. The contents are then filtered into a 100 mL volumetric flask. The solids are rinsed with water and the flask topped up to 100 mL. In order to remove the interfering chloride before the phosphate determination, 10 mL of the liquid are evaporated to dryness, the residue is dissolved in 0.5 M $HNO_3$ and the resulting solution made up to 50 mL in a volumetric flask.

### Exchangeable $Ca^{2+}$, $K^+$, $PO_4^{3-}$, $SO_4^{2-}$

Calcium, potassium and phosphate are extracted with ammonium lactate acetic acid according to the equilibrium method:

5 g air-dried fine soil are shaken for 4 hours with 100 mL of an extraction solution (consisting of 9 g lactic acid + 19 g acetic acid + 7.7 g ammonium acetate, made up to 1 L with water). The mixture is then filtered.

Sulfate is extracted with a sodium chloride solution using the equilibrium technique:

50 g air-dried fine soil is shaken with 250 mL sodium chloride solution (1 %) for 1 hour. 3 g of powdered activated charcoal are added and the mixture is shaken and filtered.

Water soluble B, $Cl^-$, $SO_4^{2-}$, $NO_3^-$, $K^+$, $Na^+$, $Ca^{2+}$, $Mg^{2+}$
_____

Boron is extracted with hot water:

25 g air-dried fine soil is boiled for 5 minutes in a flask (low in boron) with 50 mL water and then filtered.

The remaining parameters are obtained by extracting as follows:

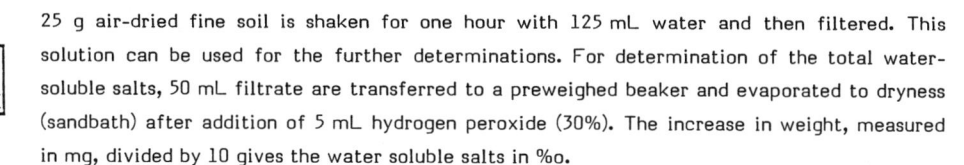

25 g air-dried fine soil is shaken for one hour with 125 mL water and then filtered. This solution can be used for the further determinations. For determination of the total water-soluble salts, 50 mL filtrate are transferred to a preweighed beaker and evaporated to dryness (sandbath) after addition of 5 mL hydrogen peroxide (30%). The increase in weight, measured in mg, divided by 10 gives the water soluble salts in %o.

## 6.2 Analytical Methods

### 6.2.1 Chemical Analyses

#### 6.2.1.1 Ammonium

Ammonium ions can be formed in water and soil by the microbiological degradation of nitrogen-containing organic compounds as well as by nitrate reduction under certain conditions. Considerable concentrations (up to 50 mg/L) are found in waste water. Very high concentrations (up to 1000 mg/L) can be encountered in seepage from refuse dumps. For this reason and within certain limits, ammonium may be regarded as a pollution indicator in ground water and drinking water.

When ammonium-containing water is brought into contact with oxygen over a long period of time, the ammonium can be microbiologically oxidized through nitrite to nitrate.

An equilibrium exists in aqueous solutions between free ammonia (toxic for fish) and ammonium ions depending on the pH value (Figure 40).

## 6 Laboratory Measurements

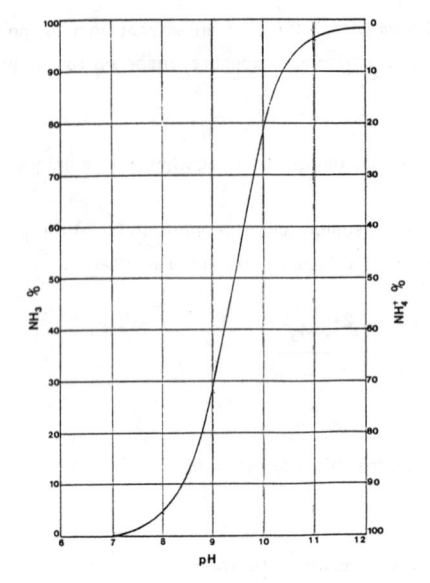

**Fig. 40:** Equilibrium between free ammonia and ammonium ions depending on pH-value

Areas of Application ⟶ water, waste water, soil

### Apparatus
Spectrophotometer or fixed filter photometer (655 nm)

Water bath

### Reagents and Solutions

Salicylate-citrate-solution: 32.5 g Na salicylate ($C_7H_5O_3Na$) and 32.5 g Trisodium citrate ($C_6H_5O_7Na_3 \cdot 2\ H_2O$) are dissolved in approx. 200 mL water. 0.243 g disodium pentacyanonitrosylferrate ($Na_2Fe(CN)_5NO \cdot 2\ H_2O$) are added and the solution made up to 250 mL. The solution lasts for about 2 weeks when stored in the dark.

Reagent solution: 3.2 g NaOH are dissolved in 50 mL water. After cooling, 0.2 g Na-dichloroisocyanurate ($C_3N_3Cl_2ONa$) are added and the solution made up to 100 mL. The solution must be freshly prepared daily.

Ammonium standard solution (1 g/L $NH_4^+$): 2.966 g ammonium chloride (dried at 105°C) are made up to 1 L with water. The dilution series for the standard curve is prepared from this solution.

## Sample Preparation

Samples containing suspended particles must be filtered (0.45 μm). Flocculation with aluminium oxide can be useful in discolored or waste waters. For this purpose, 5 to 10 mL of an aluminium sulfate solution (120 g $Al_2(SO_4)_3 \cdot 18\ H_2O$ per L) is added to 250 mL of sample followed by a $NaOH/Na_2CO_3$ solution (50 g NaOH + 50 g $Na_2CO_3$ made up with water to 300 mL) until a pH of 7 is obtained. The solution is allowed to settle and an aliquot of the clear supernatant is taken for analysis. The determination of ammonium in soil samples is carried out by alkaline distillation as described in section 6.2.1.12.

## Calibration and Measurement

Depending on the expected ammonium content, up to 40 mL of sample are pipetted into a 50 mL flask. After addition of 4 mL salicylate-citrate solution, the flask is shaken. The pH should then be 12.6 as is mainly the case for neutral waters. 4 mL of reagent solution is added, the flask is filled up, shaken and placed into a water bath at 25 °C. The measurement is carried out in a photometer (655 nm) after one hour.

The calibration series is prepared from the stock solution in the range 5 to 50 μg/40 mL sample. A blank is treated in the same way as the samples.

## Interfering Factors

The inorganic substances found at normal concentrations in water do not interfere. Even urea does not infere. However, small amine concentrations can cause problems.

## Calculation of Results

The calculations are based upon the standard curve.

### 6.2.1.2 Biochemical Oxygen Demand

Biochemical Oxygen Demand (BOD) is defined as the quantity of dissolved oxygen which is able to oxidize the organic components in the water with the assistance of microorganisms and under defined experimental conditions. The BOD is an empirical biological test in which the water conditions such as temperature, oxygen concentration or type of bacteria play a decisive role. These and other factors therefore cause the reproducibility to be much less than that of pure chemical tests. In spite of this disadvantage, the BOD is of special importance in the assessment of polluted surface waters and waste water. Its application is indispensable as in laying out data during the construction of sewage works.

The biochemical decomposition in waste waters normally takes place in two almost distinct phases. In the first phase, organic compounds are decomposed finally yielding $CO_2$ and $H_2O$. In the second phase, known as the nitrification phase, ammonium is oxidized to nitrite and then to nitrate by the action of Nitrosomas or Nitrobacter. Despite the presence of ammonium, this nitrification does not take place in every water or waste water so that the test results can be very uncertain. Nitrification is mainly observed in waste water from sewage works since here a larger number of the slowly reproducing nitrifying bacteria are already present. In many cases of waste water analysis, the nitrification can be prevented by the action of inhibitors in order to allow greater comparability between sample series. However, such inhibitors should not be employed when investigating river waters, since environmental information is required the oxygen consumption by all substances contained in the water and not only the organic components. For an assessment of BOD in connection with water characteristics see 3.2.

A reaction time of 5 days is normally used for the measurement ($BOD_5$). The dilution method is described here in which oxygen-saturated dilution water is added to the sample. Manometric measuring systems are commercially available and deliver useful results although the data from both techniques are not directly comparable. Furthermore, it is also not permissable to convert results obtained for a particular incubation period (e.g. BOD after 5 days) into results for other times.

Areas of Application ⟶ water, waste water

### Apparatus
Narrow neck glass bottles, 250 mL with ground glass stoppers
Equipment for maintaining 20 °C (water bath, incubator)

### Reagents and Solutions

Nutrient salt solutions:
a) 42.5 g potassium dihydrogenphosphate is dissolved in approx. 700 mL water, 8.8 g sodium hydroxide and 2 g ammonium sulfate are added and the solution made up to 1 L. The pH-value is adjusted to 7.2.

b) 22.5 g magnesium sulfate ($MgSO_4 \cdot 7\ H_2O$) is dissolved in 1 L of water.

c) 27.5 g calcium chloride is dissolved in 1 L of water.

d) 0.15 g ferric chloride ($FeCl_3 \cdot 6\ H_2O$) is dissolved in 1 L of water.

Dilution water: Demineralized water is employed. 1 mL of each of the nutrient solutions is added to 1 litre of water and aeration carried out for several days in the dark.

N-allyl thiourea solution: approx. 1 mg N-allyl thiourea ($C_4H_8N_2S$) is dissolved in 100 mL of water. The solution has to be freshly prepared each day.

### Sample Preparation

The pH values of acidic or alkaline reaction samples are adjusted to 7 or 8 by HCl or NaOH. Solids can be considered in the determination but mainly samples after 2 hours sedimentation or after filtration are employed. Cooled samples are warmed to room temperature before examination.

### Measurement

The water sample is left undiluted up to an expected $BOD_5$ of 6 mg/L. In the case of water free from bacteria, 5 mL of outflow from a biological sewage treatment plant or 1 mL sedimented raw waste water are added to 1 L of the dilution water. The dilution should be carried out in such a way that after 5 days incubation time at least 2 mg/L oxygen have been consumed and the remaining oxygen concentration does not fall below 2 mg/L. Since the BOD value is unknown, several different dilutions should be prepared so that at least one falls in a favorable measuring range. A previously determined COD value can be useful in selecting the dilution whereby this value is divided by 2 and a dilution taken from Table 22:

**Table 22:** Dilution series in the determination of biochemical oxygen demand

| Expected $BOD_5$ (mg/L) | mL sample made up to 1 L |
|---|---|
| to 6 | 1000 |
| 4 - 12 | 500 |
| 10 - 30 | 200 |
| 20 - 60 | 100 |
| 40 - 120 | 50 |
| 100 - 300 | 20 |
| 200 - 600 | 10 |
| 400 - 1200 | 5 |
| 1000 - 3000 | 2 |
| 2000 - 6000 | 1 |

After dilution, the sample is mixed and carefully transferred to the test bottles avoiding formation of air bubbles. The bottles are left for a short time and any air bubbles are removed by careful knocking. The ground glass stoppers are then applied without generating any air bubbles.

The oxygen concentration is immediately determined using one of the triplicate (minimum) samples according to the Winkler method or with an oxygen electrode (section 5.2.7). The remaining bottles are then left for 5 days in the dark at 20°C. After this time, the remaining oxygen is determined. A blank sample consisting of the innoculated dilution water is measured in parallel.

### Interfering Factors

An undesirable oxygen consumption via nitrification can be prevented by addition of 1 mL of an N-allyl thiourea solution. Free chlorine, present in some waste waters after chlorination reacts with organic components within about 2 hours and does not interfere. Compounds which use up oxygen without the presence of microorganisms (e.g. iron(II), sulfite or sulfide ions) are oxidized by leaving the original sample for 2 hours with occasional shaking. The presence of toxic substances can result in very low BOD values despite the presence of sufficient degradable organic materials. In such cases, a series of measurements should be carried out at greater dilutions.

Calculation of Results

The biochemical oxygen demand is given in mg/L $O_2$

$$BOD_5 \text{ (mg/L)} = \frac{A}{B} \cdot (C - D) + D$$

A = total volume after dilution (mL)
B = volume of undiluted sample (mL)
C = oxygen consumption of diluted sample after 5 days (mg/L)
D = oxygen consumption of dilution water after 5 days (mg/L)

### 6.2.1.3 Boron

Boron is found in natural unaffected waters usually only at very low concentrations. However, in domestic waste water concentrations of several mg/L are not uncommon owing to the perborate content of detergent preparations. Such small concentrations are not harmful to man but can be harmful to certain plants e.g. citrus fruits or beans when present in irrigation water.

The method of analysis using azomethine-H is described here:

Areas of Application ⟶ water, waste water, soil

Apparatus

Spectrophotometer or fixed filter photometer (414 nm)

Reagents and Solutions

Azomethine-H solution:   1 g azomethine-H sodium salt ($C_{17}H_{12}NNaO_8S_2$) and 3 g L (+) ascorbic acid ($C_6H_8O_6$) are made up to 100 mL with water. The solution is stored in a plastic bottle and is stable for approx. one week if kept in a refrigerator.

| | |
|---|---|
| Buffer solution (pH 5.9): | 25 g ammonium acetate, 25 mL water, 8 mL sulfuric acid (29%), 0.5 mL phosphoric acid (85 %), 100 mg citric acid ($C_6H_8O_7 \cdot H_2O$) and 100 mg sodium EDTA ($C_{11}H_{14}N_2Na_2O_8 \cdot 2 H_2O$) are mixed. |
| Reagent solution: | Equal volumes of azomethine-H solution and buffer solution pH 5.9 are mixed before the analysis. The solution must be stored in a refrigerator. |
| Borate standard solution (1 mg/L B as borate): | 572 mg boric acid ($H_3BO_3$) is made up to 1 L with water. 10 mL of this solution are taken and made up to 1 L with water. |

Sample Preparation

Suspended particles should be removed by filtration through a 0.45 μm membrane filter.

Calibration and Measurement

0.5 to 6 mL of the standard solution is placed in a 50 mL volumetric flask and the solution made up to approx. 25 mL. In the same way, 25 mL of sample is pipetted into a 50 mL volumetric flask. 10 mL reagent solution is added to each of the solutions and the extinction measured at 414 nm.

Interfering Factors

Iron ions at concentrations exceeding 5 mg/L can interfere.

Calculation of Results

The boron concentration of the sample is obtained by reference to the calibration curve.

6.2.1.4 Calcium and Magnesium

Calcium and magnesium ions are present in all natural waters and are often referred to as hardness factors. "Hardness" is defined as the ability of the water to cause precipitation of insoluble calcium and magnesium salts of higher fatty acids from soap solutions. In recent times, hardness is no longer referred to and total concentrations of calcium and magnesium ions are usually given instead.

In water technology, both elements play an important role in the formation of protective coatings in the mains pipes.

The complexometric method for determination of calcium and magnesium ions will be described allowing on the one hand calcium alone to be determined and on the other hand calcium and magnesium together. Magnesium is then obtained by difference. The results are most suitably presented in mmol/L.

## 94  6 Laboratory Measurements

Area of Application ⟶ water

### a) Determination of calcium

**Apparatus**
Titration equipment

**Reagents and Solutions**

EDTA solution (0.01 M): 3.725 g ethylenedinitrilo-tetraacetic acid disodium salt ($C_{10}H_{14}N_2O_8Na_2 \cdot 2\,H_2O$) are made up to 1 L with water

indicator powder: 1 g calcon carbonic acid ($C_{21}H_{14}N_2O_7S \cdot 3\,H_2O$) is ground with 99 g sodium sulfate ($Na_2SO_4$) in a mortar.

Caustic soda solution: 8 g sodium hydroxide are dissolved in 100 mL water

**Sample Preparation**
Sample preparation is not required.

**Measurement**
100 mL water sample are treated sequentially with 2 mL caustic soda solution and 0.2 g indicator powder. Titration with EDTA solution is carried out rapidly until the color changes from red to blue.

**Interfering Factors**
Barium and strontium ions are included in the results obtained. The presence of various heavy metals can cause the color change to be unclear. Interference through iron and manganese ions can be almost entirely removed by addition of 2 to 3 mL of triethanolamine.

**Calculation of Results**
The calcium concentration of the water sample is calculated as seen below:

$$Ca^{2+}\ (mmol/L) = \frac{A \cdot 100}{B}$$

A = EDTA-solution consumed (mL)
B = sample volume (mL)

### b) Sum determination of Calcium and Magnesium

**Apparatus**
Titration equipment

**Reagents and Solutions**

EDTA solution:            as described under a)

Buffer solution pH 10:    6.75 g ammonium chloride and 0.05 g EDTA disodium-magnesium salt ($C_{10}H_{12}N_2O_8Na_2Mg$) are dissolved in 57 mL ammonia (25 %) and made up to 100 mL with water.

Indicator solution:       0.5 g eriochrome black T are dissolved in 100 mL triethanolamine.

Sample Preparation
Suspended particles causing cloudiness are filtered off (0.45 μm) before the analysis.

Measurement
It is recommended that the first titration be carried out rapidly in order to avoid problems caused by carbonate precipitation during a slow titration.

100 mL sample are treated with 4 mL buffer solution and 3 drops of indicator solution. A rapid titration with EDTA is then performed until the color changes from red to blue.

The second titration is carried out by adding approx. 0.5 mL less EDTA solution to 100 mL of sample than were consumed in the first titration. Buffer solution and indicator are then added as described above and the titration performed until the color changes from red to blue.

Interfering Factors
as described under a)

Calculation of Results

$$Ca^{2+} + Mg^{2+} \text{ (mmol/L)} = \frac{A \cdot 100}{B}$$

A = volume of EDTA solution used (mL)
B = sample volume (mL)

c) Calculation of Magnesium Concentration

The magnesium concentration is obtained by subtracting the result of measurement a) from that of measurement b).

6.2.1.5 Chemical Oxygen Demand ($KMnO_4$ consumption and oxidation with $K_2Cr_2O_7$)

Chemical oxygen demand (COD) is defined as the amount of oxygen in the form of oxidizing agent consumed in the oxidation of organic water components. The degree of oxidation depends upon the type of substance, pH value, temperature, reaction time and concentration of oxidizing agent as well as the type of added accelerators, if any.

Potassium permanganate has long been used as an oxidizing agent for determination of organic compounds in water and waste water. The procedure is relatively easy to carry out but suffers from the disadvantage that certain substances such as some amino acids, ketones or saturated carboxylic acids are not or only partially oxidized. Therefore, the method is mainly applied to lesser polluted surface waters or drinking water for survey purposes, the results being used only for orientation. The measurement is mainly performed in acidic solution in which the permanganate ion is reduced to Mn(II).

$$MnO_4^- + 8\,H^+ + 5\,e^- \rightleftharpoons Mn^{2+} + 4\,H_2O$$

More accurate COD values are given by oxidation through potassium dichromate in strongly acidic solution.

$$Cr_2O_7^{2-} + 14\,H^+ + 6\,e^- \rightleftharpoons 2\,Cr^{3+} + 7\,H_2O$$

This method is used for determination of the COD in all water and waste water types. With few exceptions, all organic substances are almost completely oxidized. Concentrations (in mg/L $O_2$) from 10 to 15 mg/L can normally be safely measured. A modified method is employed for smaller concentrations (Literature list: Deutsche Einheitsverfahren).

For the interpretation of results, it is important to note that the COD value cannot be directly converted into a measure of the amount of organic substances present where the quantitative composition is unknown. Different substances require different amounts of oxidizing agent for complete oxidation, e.g.:

| | | |
|---|---|---|
| oxalic acid | $(C_2H_2O_4)$ | 0.18 mg/mg substance |
| acetic acid | $(C_2H_4O_2)$ | 1.07 mg/mg substance |
| phenol | $(C_6H_6O)$ | 2.38 mg/mg substance |

For domestic waste water, a value of 1.2 mg COD per mg organic material may be assumed.

In addition to organic substances, various inorganic ions can be simultaneously oxidized (e.g. nitrite, sulfite, Fe(II)). Higher chloride concentrations can interfere and are therefore masked with mercury ions or degassed as HCl before the determination. Silver ions are added to accelerate the oxidation.

<u>Area of Application</u> ⟶ water, waste water, soil ($K_2Cr_2O_7$ method)

a) <u>Potassium permanganate consumption</u>

<u>Apparatus</u>

Titration equipment

Gas burner or heating plate

## Reagents and Solutions

Potassium permanganate solution:
3.1608 g potassium permanganate is made up to 1 L with freshly boiled and cooled distilled water. 100 mL is taken and made up to 1 L. The titration factor is always determined with oxalic acid before use. The solutions are stored in dark bottles.

Oxalic Acid:
6.3033 g oxalic acid ($C_2H_2O_4 \cdot 2\ H_2O$) is made up to 1 L with freshly distilled water and careful addition of 50 mL concentrated sulfuric acid. 100 mL of this solution is made up to 1 L with water and careful addition of 50 mL conc. sulfuric acid. The solution is stable for up to 6 months in dark bottles.

Sulfuric Acid (d = 1.27):
100 mL concentrated sulfuric are carefully added to 200 mL water. $KMnO_4$ solution is added to the solution while hot until a pink color appears.

## Sample Preparation

The measurement should be carried out as soon as possible after sample collection. The glass apparatus employed for boiling should be stored under dust-free conditions.

## Measurement

The adjustment of the titration factor of the $KMnO_4$-solution is carried out by heating to boiling 100 mL distilled water after adding anti-bump granules and 5 mL sulfuric acid. $KMnO_4$ solution is added until a weakly pink color appears. Then 20 mL oxalic acid are added from a pipette and the solution again titrated with $KMnO_4$ until weakly pink. The consumption x should lie between 19 and 21 mL. The factor is determined as $f = 20/x$.

The measurement is performed by adding 100 mL of sample (or a smaller volume made up to 100 mL) to an Erlenmeyer flask followed by 5 mL sulfuric acid (d = 1.27). The flask is covered with a cold finger or watch glass and brought to boiling within 5 minutes. While boiling, 20 mL of potassium permanganate solution is added from a pipette and the solution allowed to simmer for 10 minutes. Then, 20 mL of oxalic acid solution are added and the mixture heated until the color completely disappears. The approx. 80 °C hot solution is titrated with potassium permanganate solution until a pink color lasts for approx. 30 seconds. The volume consumed should lie between 4 and 12 mL. Where the consumption is found to be higher or the sample already colorless before addition of oxalic acid, the determination is repeated with a smaller sample volume. A blank with 100 mL of dilution water is measured in parallel.

## Interfering Factors

Hydrogen sulfide, sulfides and nitrites interfere but are removed in an acidic environment. Chloride ion concentrations exceeding 300 mg/L can interfere. In such cases, the sample can be diluted to obtain a concentration below this value.

### Calculation of Results

The KMnO$_4$-consumption is given in mg/L KMnO$_4$

$$KMnO_4 \text{ (mg/L)} = \frac{(a-b) \cdot f \cdot 316 \text{ mg}}{V}$$

a  KMnO$_4$ consumed by sample (mL)
b  KMnO$_4$ consumed by blank (mL)
f  titration factor of the KMnO$_4$ solution
V  sample volume (mL)

### b) Potassium dichromate consumption

#### Apparatus
Reflux equipment consisting of a 250 mL flask (Erlenmeyer or round-bottom) with ground glass neck and reflux condenser
Heating mantle or plate
Titration equipment

#### Reagents and Solutions

Sulfuric acid, containing silver: 15 g silver sulfate are dissolved in 1 L concentrated sulfuric acid.

Potassium dichromate solutions: 12.259 g potassium dichromate (dried for 2 hours at 105 °C) are made up to 1 L with water.

100 mL of this solution are made up to 1 L with water.

Ferrous ammonium-sulfate solutions: 98 g ferrous ammonium sulfate (Fe(NH$_4$)$_2$(SO$_4$)$_2$ · 6 H$_2$O) are dissolved in water. 20 mL of concentrated sulfuric acid are added and the solution made up to 1 L with water. 100 mL of this solution is made up to 1 L with water.

The factors of these solutions are adjusted with K$_2$Cr$_2$O$_7$ solution before use.

Ferroin indicator solution: 0.98 g ferrous ammonium sulfate solution and 1.485 g 1,10-phenanthroline (C$_{12}$H$_8$N$_2$ · H$_2$O) are made up to 100 mL with water.

Mercurous sulfate solution: For masking of chloride concentrations of 500 or 1500 mg/L, 10 mL concentrated sulfuric acid and 5 g or 15 g mercurous sulfate (HgSO$_4$) are made up to 100 mL with water.

### Sample Preparation

The measurement should be performed as soon as possible after sample collection. Glass equipment should be kept under dust-free conditions and ground-glass joints must not be greased. Waste water samples are normally tested after being left for 2 hours to remove solids by sedimentation.

### Measurement

The factor adjustment of the ferrous ammonium sulfate solution is carried out as follows: 10 mL of the more concentrated $K_2Cr_2O_7$ solution diluted to 100 mL. 30 mL concentrated sulfuric acid added, the solution is cooled and titrated with ferrous ammonium sulfate after addition of 3 drops of ferroin indicator. The amount x consumed (mL) is used to calculate the factor according to $f = 10/x$, which must be checked daily.

The determination is carried out as follows: 20 mL sample (or an aliquot diluted to 20 mL) and 10 mL of the more concentrated dichromate solution are placed into the flask together with anti-bump granules. Afterwards 30 mL of the silver-containing sulfuric acid are carefully added with shaking. The solution is then carefully boiled under reflux for 2 hours.

Where preliminary measurements on known water or waste water samples have shown that a shorter reflux time gives the same results, the shorter time should be used.

After heating, the solution is cooled and the condenser rinsed with distilled water until the volume of solution in the flask is about 150 mL. After cooling to room temperature, the excess of dichromate is back-titrated with the ferrous ammonium sulfate solution. The indicator changes color from blue-green to red-brown at the end-point. A control sample consisting of 20 mL distilled water is measured at the same time.

Addition of mercurous sulfate solution is not necessary where the chloride concentration is less than 100 mg/L. At concentrations of 500 or 1500 mg/L chloride, 2 mL of the respective mercurous sulfate solution are added before refluxing.

The less concentrated dichromate and ferrous ammonium sulfate solutions are employed for very low concentrations of organic substances. All chemicals should be of the highest purity. Glass equipment is thoroughly cleaned with chromic sulfuric acid before use. The quality of the method may be tested by potassium phthalate. 425.1 mg potassium phthalate (dried at 105 °C) are made up to 1 L. This solution has a COD value of 500 mg/L.

### Interfering Factors

Interference caused by chloride ions can be prevented as described above.

### Calculation of Results

The COD value is given in mg/L $O_2$.

$$\text{COD (mg/L)} = \frac{(a-b) \cdot f \cdot 2000\,mg}{V}$$

a = ferrous ammonium sulfate consumed by blank (mL)
b = ferrous ammonium sulfate consumed by sample (mL)
f = titration factor of the ferrous ammonium-sulfate solution
V = sample volume (mL)

### 6.2.1.6 Chloride

Chloride is present in all natural waters at greatly varying concentrations depending on the geochemical conditions. Particularly high concentrations occur in waters near salt deposits. Large amounts of chloride reach waste water through faecal introduction. For this reason, chloride can serve as a pollution indicator when considered together with other parameters and when a natural geological origin does not apply. A concentration of even 250 mg/L can cause a salty taste in some waters whereas in other waters containing larger amounts of calcium and magnesium ions, a concentration of approx. 1000 mg/L is necessary.

The precipitation titration with mercurous ions is suitable but use of mercury is to be discouraged owing to waste disposal problems and toxicity. The silver nitrate titration with potassium chromate as end-point indicator is described below:

Areas of Application ⟶  Water, waste water, soil

Apparatus
Titration equipment

Reagents and Solutions

Silver standard solution: 4.791 g silver nitrate is dissolved and made up to 1 L with water. The solution is stored in a brown glass bottle. 1 mL corresponds to 1 mg chloride.

Sodium chloride standard solution: 1.648 g sodium chloride (dried for 2 hours at 105 °C) is dissolved and made up to 1 L with water. 1 mL of this solution contains 1 mg chloride.

Potassium chromate solution: 10 g potassium chromate is dissolved and made up to 100 mL with water.

Sample Preparation
At pH values below 5, a small amount of calcium carbonate is added and the sample shaken. At pH values exceeding 9.5, the sample is first of all titrated against phenolphthalein with 0.1 M

$H_2SO_4$. The acid amount obtained is then added to a second sample together with some calcium carbonate.

### Measurement

100 mL of sample (or a smaller volume at higher chloride concentrations) are placed into an Erlenmeyer flask and 1 mL potassium chromate solution added. The mixture is then titrated against a white background with silver nitrate solution until the color changes from greenish-yellow to reddish-brown.

A blank sample with distilled water is treated in the same way as the sample.

### Interfering Factors

Bromide, iodide or cyanide are also included in the measurement. Sulfite, sulfide and thiosulfate can be removed as follows:

After careful acidification of the sample with sulfuric acid, the mixture is boiled for several minutes, 3 mL $H_2O_2$ (10 %) is added and the boiling continued for another 15 minutes. The evaporative losses are then compensated. 1 M NaOH solution is added dropwise until weakly alkaline and the mixture boiled for a short time again. After filtration, the analysis is performed as described above.

Discolored water samples can be treated with activated charcoal followed by filtration.

### Calculation of Results

$$\text{Chloride concentration (mg/L)} = \frac{(A - B) \cdot 1000 \text{ mg}}{C}$$

A = volume of $AgNO_3$ solution consumed (mL) for sample
B = volume of $AgNO_3$ solution (mL) for blank
C = sample volume (mL)

### 6.2.1.7 Copper

In natural unaffected waters, the concentration of copper does not exceed a few µg/L. Low concentrations (0.1 to 0.2 mg/L) in contaminated water systems can be toxic towards lesser water organisms. Higher concentrations in drinking water are generally a result of corrosion in copper pipes and values of up to 3 mg/L are not uncommon after a long standing time.

A relatively simple photometric method is described below where interference problems are greater than with the atomic absorption spectrometry method requiring more complex equipment.

Area of Application ⟶  Water

### Apparatus
Spectrophotometer or fixed filter photometer (440 nm)
Separating funnel 250 mL

### Reagents and Solutions

DDTC Solution: 1 g sodium diethyldithio carbaminate ($C_5H_{10}NS_2Na$) are dissolved in 100 mL water. The solution is stable for only a few days.

Citric acid solution, 20 %

Sulfuric acid, 35 %

Ammonium chloride solution, 20 %

Chloroform

Copper standard solution (10 mg/L): 1 g metallic copper is dissolved in 10 mL nitric acid (30 %) and made up to 1 L with water. 10 mL are taken and made up to 1 L with water.

### Sample Preparation
No sample preparation is required for clear solutions.

### Calibration and Measurement
Volumes of 0 to 30 mL copper standard solution (corresponding to 0 to 0.3 mg Cu) are treated in the same way as the sample.

100 mL sample (if necessary made up to 100 mL) are transferred to a separating funnel and the following reagents are added: 1 mL citric acid solution (20 %), 0.5 mL sulfuric acid (35 %), 0,5 mL ammonium chloride solution (20 %) and 10 mL chloroform. After shaking for some minutes, the organic phase is discarded. 2 mL DDTC solution are added together with 25 mL chloroform and shaking continued for 3 minutes. The organic phase is then filtered and measured at 440 nm in the photometer.

### Interfering Factors
Higher concentrations of zinc and heavy metals can interfere. The copper complex extracted in the chloroform is light sensitive so that measurement must be carried out immediately after extraction.

### Calculation of Results
The copper content is obtained by reference to the calibration curve.

### 6.2.1.8 Cyanides

Cyanides are employed in many industrial and manufacturing areas and can reach water systems via waste water pipes. They occur as cyanide ions, hydrogen cyanide or as complex bound cyanides.

Cyanide ions are toxic towards water organisms whereas many complexed cyanides are only minimally toxic. Such complex cyanides can decompose or rearrange under certain conditions, e.g. depending on pH, to yield cyanide ions so that an assessment of the toxic properties of water samples is not always a simple matter.

In order to cover the various forms of cyanide compounds, a conventional distinction is made between "total cyanide" and "easily released cyanide". Total cyanide is the sum of simple and complex cyanides whereas the easily released cyanide includes those compounds which release hydrogen cyanide at room temperature and pH 4.

The presence of cyanides in ground water, surface water and drinking water points to the influence of waste or waste water and always requires counter-measures. Raw water used in drinking water processing and drinking water itself must not contain cyanide concentrations in excess of 0.05 mg/L.

Areas of Application ⟶ Water, waste water

## a) Total Cyanides

### Apparatus
Distillation equipment with 3-neck flask, heater, reflux condenser, absorption container, funnel, water pump, wash bottle, flow meter (Figure 41)
Spectraphotometer or fixed filter spectrometer, 580 nm.

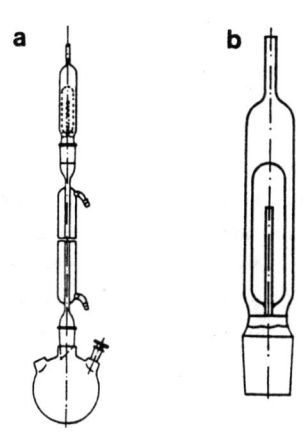

**Fig. 41:** Distillation apparatus (a), absorption tube (b)

### Reagents and Solutions
Hydrochloric acid, 25 %
Hydrochloric acid, 1 M
Sodium hydroxide solution, 20 %

Sodium hydroxide solution, 1 M

Stannous chloride solution: 50 g stannous chloride ($SnCl_2 \cdot 2 H_2O$) is dissolved in 40 mL hydrochloric acid (1 M) and made up to 100 mL with water.

Chloroform-phenolphthalein solution: 0.03 g phenolphthalein is dissolved in 90 mL ethanol and then 10 mL chloroform added.

Zinc-cadmium-sulfate solution: 10 g zinc sulfate ($ZnSO_4 \cdot 7 H_2O$) and 10 g cadmium sulfate ($3 CdSO_4 \cdot 8 H_2O$) are dissolved and made up to 100 mL with water.

Buffer solution pH 5.4: 60 g sodium hydroxide is dissolved in 500 mL water and treated with 118 g succinic acid. The solution is made up to 1 L with water.

Chloramine-T solution: 1 g chloramine-T ($C_7H_7SO_2NClNa$) is dissolved in 100 mL water. The solution is stable for approx. 1 week.

Barbituric acid-pyridine reagent: 3 g barbituric acid ($C_4H_4O_3N_2$) is placed into a 50 mL volumetric flask, mixed with some water and treated with 15 mL pyridine. Sufficient water is added with shaking to completely dissolve the barbituric acid. 3 mL hydrochloric acid (25 %) is then added and the flask filled with water. The reagent is stable for one week when stored in the refrigerator.

Cyanide standard solution (10 mg/L $CN^-$): 25 mg potassium cyanide is dissolved in 1 L of 1 M sodium hydroxide solution.

Sample Preparation

After collection, 5 mL of 1 M sodium hydroxide, 10 mL chloroform-phenolphthalein solution and 5 mL stannous chloride solution are added to 1 L of sample. In the case of reddening, 1 M hydrochloric acid is added dropwise until colorless. Where no color is obtained, 1 M sodium hydroxide is added dropwise until red. 10 mL zinc-cadmium sulfate solution is then added and the sample stored under cool, dark conditions until analysis. The container must be shaken well before taking an aliquot for analysis.

Calibration and Measurement

100 to 400 mL sample are placed in the distillation apparatus. 10 mL of 1 M sodium hydroxide solution is placed into the absorption container and aerated using an air stream of approx. 20 L/h. The air is previously drawn through a wash bottle containing 100 mL of 1 M sodium hydroxide.

10 mL hydrochloric acid (25 %) is added to the sample and the mixture gently boiled for

approx. 45 minutes. For determination of cyanide ions, the content of the absorption container is then transferred with washing to a 25 mL volumetric flask which is then filled with water. 10 mL are removed from this flask and placed in another 25 mL volumetric flask. 2 mL buffer solution, 4 mL of 1 M hydrochloric acid and 1 mL chloramine T solution are added and the mixture left for between 1 and 5 minutes. 3 mL barbituric acid-pyridine reagent are then added and the flask is topped up with water. The contents are measured photometrically at 580 nm after 20 minutes against a control solution containing the same concentrations of the substances mentioned above. A blank sample of distilled water is treated in the same manner as the sample.

Interfering Factors

Sulfur dioxide or nitrogen oxides are not permitted to enter the absorption liquid since they interfere with the measurement. Similarly, compounds such as aldehydes which affect the reaction of chloramine T interfere.

Calculation of Results

$$\text{Total cyanides (mg/L)} = \frac{(A - B) \cdot 2500 \text{ mg}}{e \cdot V}$$

A = cyanide content of sample (mg)
B = cyanide content of blank (mg)
e = factor 0.97 accounting for volume increase by addition of preservative
V = sample volume (mL)

b) easily released cyanide

Apparatus

Distillation equipment, as described under a)

Reagents and Solutions

Hydrochloric acid, 36 %

Buffer solution, pH 4: 80 g potassium hydrogen phthalate ($C_8H_5O_4K$) is dissolved in 920 mL warm water

EDTA solution: 100 g sodium ethylene dinitrolotetra acetate ($C_{10}H_{14}N_2O_8Na_2 \cdot 2 H_2O$) is dissolved in 940 mL warm water.

Zinc powder

The further required solutions are as listed under a).

Sample Preparation

as described under a)

### Calibration and Measurement

10 mL of 1 M sodium hydroxide solution are placed into the absorption container and the air current adjusted to 30 to 60 L/h. 10 mL zinc cadmium sulfate solution, 10 mL EDTA solution, 50 mL buffer solution and 100 mL to 200 mL of the well shaken sample are then added. The pH is adjusted to 4 with NaOH or HCl and 0.3 g zinc powder added. The flask is then connected with a wash bottle (filled with 100 mL of 1 M sodium hydroxide) and the air current adjusted to 60 L/h. A blank is treated in the same way as the sample.

The calibration, measurement and calculation of results are as described under a), above.

### 6.2.1.9 Dissolved and Undissolved Substances

Dissolved and undissolved substances in water are separated by filtration. Filterable substances are designated as those undissolved substances which are separable by a medium coarse filter and weighed after drying for 2 hours at 105 °C.

The evaporation residue includes the organic and inorganic substances which are involatile at temperatures up to 105 °C.

The ash residue includes those substances which are weighed after 1 hour heating at red hot temperatures (600 °C to 650 °C). The loss on ashing is given as the difference between the evaporation residue and the ash value. It arises from organic components as well as the decomposition of inorganic salts such as nitrates, carbonates or hydrogen carbonates.

The total salt content of water can be determined by passing the sample through a cation exchanger followed by acid titration.

Areas of Application ⟶ Water, waste water, soil (dry residue, ash)

#### a) Filterable Substances

##### Apparatus
Funnels
Filter papers, mid coarse or G 2 sintered filters

##### Measurement
For filtration purposes, sufficient sample is used so that at least 10 mg solids are collected. If necessary, the filtration is carried out under suction and includes washing with distilled water. The filter paper is then dried in a weighing glass at 105 °C for 2 hours, allowed to cool in a desiccator and weighed.

## Calculation of Results

Filterable substances (mg/L) = $\dfrac{m \cdot 1000}{V}$

m = weight of residue (mg)
V = sample volume (mL)

### b) Evaporation Residue

**Apparatus**
Water bath or laboratory radiation heater
Porcelain crucibles (15 cm diameter)

**Measurement**
Following filtering through a medium coarse filter paper, the water sample (at least 100 mL) is placed in a predried and preweighed porcelain crucible and evaporated to dryness on the water bath or under the radiator. The evaporation can be carried out in stages, the sample being dried to constant weight. Soil samples are measured by placing 5 to 10 g of air-dried fine soil into a porcelain crucible and drying at 105 °C to constant weight.

**Calculation of Results**

Evaporation residue (mg/L) = $\dfrac{m \cdot 1000}{V}$

m = weight of residue (mg)
V = sample volume (mL)

### c) Ash residue

**Apparatus**
Porcelain crucibles (15 cm diameter)
Muffle oven

**Reagents and Solutions**
Ammonium nitrate solution: 1 g ammonium nitrate is dissolved in 100 mL water.

**Measurement**
The porcelain crucible containing the evaporation residue is heated for 1 hour at 600 °C. The crucible should have previously been heated to 600 °C and weighed. Where black residues are found after heating, they are moistened with several drops of the $NH_4NO_3$-solution after cooling. The residue is then carefully dried and heated for a further 10 minutes at 600 °C. The basin is then weighed after drying in a desiccator.

For soil samples, the dry residue is heated in a porcelain crucible at 600 °C to constant weight.

## Calculation of Results

$$\text{Ash} = \frac{m \cdot 1000}{V}$$

$$\text{Heating losses} = \frac{(m - m_1) \cdot 1000 \text{ mg}}{V}$$

$m_1$ = weight of ash residue (mg)
$m$ = weight of evaporation residue (mg)
$V$ = sample volume (mL)

### d) Total salt content

**Apparatus**

Glass column (approx. 1 cm diameter) with tap
Glass wool
pH paper
Titration equipment

**Reagents and Solutions**

Strongly acidic cation exchanger (e.g. Dowex 50 W)
Sodium hydroxide solution, 0.1 M
Methyl red indicator solution
Sulfuric acid, 12 %

**Measurement**

Approx. 5 mL cation exchanger resin is allowed to swell for 30 minutes in water and carefully filled into the column (regeneration is carried out by passing approx. of 20 mL 12 % sulfuric acid at a rate of 1 mL/minute through the column. Washing is carried out with water until the pH of the water used is obtained).

The measurement is performed by passing approx. 50 mL of the filtered water sample through the column at a rate of 1 mL/minute. After washing the column twice with 10 mL water, the percolate is titrated with 0.1 M NaOH against methyl red. The resin capacity must be considered in the measurement. The saturation point lies around 0.9 milligramme equivalents per mL of exchange resin.

**Calculation of Results**

The result can be calculated in terms of the monovalent $Na^+$ cation

$$Na^+ \text{ (mg/L)} = \frac{V_1 \cdot 2.23 \cdot 1000 \text{ mg} \cdot mL^{-1}}{V}$$

$V_1$ = volume of 0.1 M NaOH consumed (mL)
$V$ = sample volume (mL)
1 mg $Na^+$ ≙ 2,62 mg NaCl

## 6.2.1.10 Iron

Iron is present at different concentrations in many waters and waste waters. Under anaerobic conditions, several mg/L iron in the form of $Fe^{2+}$ can be present whereas concentrations in aerobic surface waters seldom exceed 0.3 mg/L. $Fe^{2+}$ ions can be fairly rapidly oxidized by atmospheric oxygen. First, yellow-brown colloidal ferric hydroxide is formed which then precipitates as brown hydroxide.

$$2\, Fe^{2+} + 1/2\, O_2 + 2\, H^+ \rightarrow 2\, Fe^{3+} + H_2O$$

This process plays an important role in drinking water technology. Iron is an undesirable component of drinking and industrial water because iron hydroxide can form deposits in pipes or cause problems in usage (e.g. metallic taste, stains in textiles after washing).

In the determination of iron, a distinction is made between total iron and ferrous ions. The method described below regarding the analysis of iron uses 2,2'-dipyridyl. The total iron is the sum of dissolved and undissolved iron and is distinguished from the total dissolved iron which is the sum of ferrous and ferric compounds.

Areas of Application $\longrightarrow$ Water, waste water

### a) Ferrous (iron-II)

**Apparatus**
Spectrophotometer or fixed filter photometer (520 nm)

**Reagents and Solutions**

2,2'-dipyridyl solution: 0.1 g 2,2'-dipyridyl ($C_{10}H_8N_2$) is dissolved and made up to 100 mL with water.

Ferrous standard solution (100 mg/L, 10 mg/L): 0.7022 g ferrous ammonium sulfate (($NH_4$)$_2$Fe($SO_4$)$_2 \cdot 6\, H_2O$) is dissolved and made up to 1 L with water. 100 mL of this solution is made up with water to 1 L. Both solutions must be freshly prepared.

**Sample Preparation**
20 mL of the dipyridyl solution are placed into 100 mL graduated flasks in the laboratory prior to sample collection. 10, 20, 50 or 75 mL of sample are then pipetted during collection into these flasks depending on the expected $Fe^{2+}$ concentration.

**Calibration and Measurement**
Volumes of the fresh standard solutions corresponding to between 0.05 and 1 mg/L $Fe^{2+}$ are pipetted into 100 mL volumetric flasks containing 20 mL 2,2'-dipyridyl solution. After filling

up the flasks with water, measurements are made in the photometer at 520 nm.

$Fe^{2+}$ in the samples is determined in the same manner after choosing a suitable sample volume. Where the pH value of the solution lies outside the range 3 - 9, a small amount of sodium acetate buffer is added. A blank is measured in parallel.

Interfering Factors
Turbid solutions should be rapidly filtered before analysis.

Calculation of Results
The concentration of $Fe^{2+}$ ions is calculated by reference to the standard curve and the measured blank.

b) Total Iron

Apparatus
as described under "Ferrous"

Reagents and Solutions
2,2'-dipyridyl solution:    as described under "Ferrous"

Hydrochloric acid, 25 %
Nitric acid, 65 %
Sodium hydroxide, 5 %
Sodium acetate ($CH_3COONa \cdot 3 H_2O$), solid
Ascorbic acid ($C_6H_8O_4$), solid

Ferric standard solution (100 mg/L, 10 mg/L):    100 mg iron wire is dissolved with warming in a 1 L volumetric flask containing 20 mL $H_2O$ and 5 mL hydrochloric acid (36%). After cooling, the content is made up to 1 L. 100 mL of this solution is made up to 1 L with water. The diluted solution should be freshly prepared.

Sample Preparation
Samples for total iron determination are acidified to pH 1 with hydrochloric acid during collection. For determination of total dissolved iron, the samples are filtered through a medium coarse filter paper immediately after sampling and acidified as above.

Dissociation is necessary for the inclusion of sparely soluble or insoluble iron compounds. 50 mL sample are placed in a beaker together with 5 mL nitric acid (65 %) and 10 mL hydrochloric acid (25 %) and heated to about 80 °C until all solids are dissolved. Then 2 mL of concentrated sulfuric acid are added and the solution evaporated down until white $SO_3^-$ fumes appear. After cooling, 20 mL water are added, the solution is filtered if necessary and made up

### Calibration and Measurement

Volumes of the standard solutions containing amounts of iron between 0.05 and 1 mg/L are taken and treated like the samples.

The total iron is determined as follows:
50 mL (or a smaller volume) of sample is pipetted into a 100 mL volumetric flask and if necessary, the pH adjusted to a value between 2 and 3 with sodium hydroxide (5 %). After adding 300 to 500 mg ascorbic acid to reduce $Fe^{3+}$ to $Fe^{2+}$, 20 mL 2,2'-dipyridyl solution are added. The mixture in buffered to pH 5 - 6 by addition of 2 to 5 g solid sodium acetate and made up to 1 L with water. Photometric measurements are then performed at 520 nm. A blank is treated in exactly the same way as the sample.

### Interfering Factors

Copper, zinc and o-phosphate can interfere when present at concentrations exceeding that of iron by about a factor 10.

Interfering cyanides are normally removed by the addition of acid. Where the presence of bound cyanide complexes is suspected, the dissociation method should be employed.

### Calculation of Results

The total iron concentration is calculated by reference to the standard curve and measured blank.

### 6.2.1.11 Kjeldahl Nitrogen

Kjeldahl nitrogen is the sum of ammonium and those organic nitrogen compounds which can be converted to ammonium under Kjeldahl reaction conditions. The portion of organic nitrogen may be obtained by subtracting the ammonium content from the Kjeldahl nitrogen value.

Low concentrations of soluble organic nitrogen compounds can be present in natural waters as a result of organic compound degradation. Further enzymatic conversion to ammonium is possible. Waste water, however, generally contains higher concentrations of organic nitrogen compounds.

### Areas of Application ⟶ Water, waste water, soil

### Apparatus
Kjeldahl flasks, 350 mL
Distillation apparatus with 1 L flasks, Liebig condenser, heater

### Reagents and Solutions

## 6 Laboratory Measurements

Sodium hydroxide solution: 400 g sodium hydroxide are dissolved in 1 L of water.

Reaction mixture: 5 g selenium, 5 g copper sulfate ($CuSO_4$), 250 g sodium sulfate (anhydrous) are mixed in a mortar and stored under dry conditions.

Sulfuric acid, concentrated
Sulfuric acid, 0.025 M

Phenolphthalein solution: 1 g phenolphthalein is dissolved in 100 mL ethanol. 100 mL water are then added.

Mixed indicator solution:
a) 30 mg methyl red is dissolved in 100 mL ethanol,
b) 100 mg methylene blue is dissolved in 100 mL water,
100 mL solution a) is mixed with 15 mL solution b).

Ethanol

### Sample Preparation

Sample preparation is not necessary. In the case of water samples the determination should be carried out promptly after collection. In the case of soil, 1 to 5 g of air-dried soil is used.

### Measurement

A 100 mL water sample is placed into the Kjeldahl flask and treated with 1 g of reaction mixture and 10 mL of ethanol. After shaking, 10 mL of conc. sulfuric acid is added and the mixture heated to boiling until a light green color is obtained and black particles are no longer visible. Boiling is then continued for 30 minutes. This process removes nitrite and nitrate. After cooling and diluting with water to a total volume of approx. 300 mL, the content is transferred to the 1 L flask after rinsing the Kjeldahl flask twice. A few drops of phenolphthalein solution are then added together with sufficient sodium hydroxide to color the content red.

The flask is then attached to the distillation apparatus and approx. 200 mL distilled over, during which the end of the condenser should dip into the absorber. The ammonium content is then determined titrimetrically or photometrically depending on the original nitrogen content. The photometric method is preferred for a nitrogen content of less than 100 mg/L whereas the titrimetric method is employed at higher concentrations.

### Titrimetric Determination

The distillate is collected in a 250 mL volumetric flask containing 50 mL water. 100 mL of this solution are mixed with 3 drops of the mixed indicator solution and the titration carried out with 0.025 M sulfuric acid until a color change from violet to green is observed. A blank sample of water is titrated in the same way.

### Photometric Determination

50 mL of the distillate made up to 250 mL are analyzed as described under "Ammonium" (section 6.2.1.1).

### Interfering Factors

Various aromatic and heterocyclic nitrogen compounds are not completely covered by the method.

### Calculation of Results

Titrimetric determination

$$\text{Kjeldahl N (mg/L)} = \frac{a \cdot b \cdot 700}{c \cdot d} \frac{mg}{mL}$$

a = consumption of 0.025 M $H_2SO_4$ (mL)
b = aliquot of distillate taken (mL)
c = sample volume (mL)
d = total volume of distillate (mL)

Photometric determination:
Results are calculated by reference to a calibration curve.

### 6.2.1.12 Manganese

Manganese is often found in clean surface waters at concentrations of several tenths mg/L. However, levels exceeding 1 mg/L can occur under anaerobic conditions. As is the case with iron (together with which this element is often found) the presence of manganese is generally undesirable in the water mains owing to occurrence of oxidative precipitates. In addition, even small concentrations can be detrimental to taste. Removal of manganese from raw water for drinking and industrial use presents greater technical problems than the removal of iron. The water is aerated and passed through a gravel filter which has a manganese oxide surface already formed on it.

The analysis of manganese can be carried out by the formaldoxime reaction or through oxidation to the permanganate ion by peroxodisulfate. The former method is preferred for relatively clean waters and the latter for polluted and discolored waters.

Areas of Application ⟶ Water, waste water

#### a) Determination with formaldoxime

### Apparatus

Spectrophotometer or fixed filter photometer (480 nm)

### Reagents and Solutions

**Formaldoxime solution:** 4 g hydroxylammonium chloride ($NH_2OH \cdot HCl$) and 0.8 g paraformaldehyde ($CHOH_x$) is made up to 100 mL with water.

**Ferrous ammonium sulfate solution:** 140 g ferrous ammonium sulfate $(NH_4)_2 Fe(SO_4)_2 \cdot 6 H_2O$ is treated with 1 mL concentrated sulfuric acid and made up to 100 mL with water.

**EDTA solution:** 40 g ethylene dinitrilo tetraacetic acid sodium salt ($C_{10}H_{14}O_8N_2Na_2 \cdot 2 H_2O$) is dissolved in 100 mL water.

**Hydroxylammonium chloride solution:** 10 g hydroxylammonium chloride ($NH_2OH \cdot HCl$) is made up to 100 mL with water.

**Ammonium solution:** 75 mL of ammonium hydroxide (25 %) is mixed with 25 mL of water.

**Manganese standard solutions (100 mg/L, 10 mg/L):** 308 mg manganese sulfate ($MnSO_4 \cdot H_2O$) is dissolved in 100 mL water, 3 mL of concentrated sulfuric acid is carefully added and the solution is made up to 1 L. 100 mL of this solution is made up to 1 L with water.

### Sample Preparation

Calcium and magnesium concentrations in excess of 300 mg/L lead to artificially high values so that dilution is necessary in such cases.

Suspended particles are removed by centrifugation prior to the photometric determination.

### Measurement

The following reagents are added sequentially to 50 mL water sample with shaking: 5 mL formaldoxime solution, 5 mL ferrous ammonium sulfate solution, 5 mL ammonia solution, and after 5 minutes 5 mL EDTA solution and 5 mL hydroxylammonium chloride solution. The mixture is measured photometrically at 480 nm after a least one hour. A blank is treated in an identical manner.

### Interfering Factors

Interference caused by ferrous ions is prevented by addition of EDTA and hydroxylammonium chloride.

Phosphate concentrations exceeding 10 mg/L together with the presence of calcium ions can result in too small values being obtained.

Discolored samples are analyzed by the method described below.

## Calculation of Results
Results are calculated by reference to a calibration chart.

## b) Determination as Permanganate

### Apparatus
Spectrophotometer or fixed filter photometer (525 nm)

### Reagents and Solution
Ammonium peroxodisulfate $(NH_4)_2S_2O_8$

Reaction solution: 7.5 g mercury sulfate $(HgSO_4)$ is dissolved in 40 mL nitric acid (65%) and 20 mL water. 20 mL orthophosphoric acid (85%) are added together with 3.5 mg silver nitrate and the mixture made up to 100 mL with water.

Manganese standard solution: as described under a) above.

### Sample Preparation
The sample should be acidified after collection in order to prevent precipitation of insoluble manganese compounds. Since the permanganate ions are sensitive to the presence of reducing substances, these must be removed before analysis. Materials causing turbidity are filtered off and organic substances (from approx. 60 mg/L $KMnO_4$ consumption) are removed by oxidation with $HNO_3$: 100 mL sample containing 1 mL concentrated $H_2SO_4$ and 1 mL $HNO_3$ (65 %) are evaporated until white $SO_3$-fumes are visible. In cases of brown discoloration, some water is added and small amounts of $HNO_3$ are given repeatedly.

The residue is then mixed with diluted $HNO_3$ and made up to 100 mL with water.

### Measurement
5 mL reaction solution and 1 g solid $(NH_4)_2 S_2O_8$ are added to 90 mL of the (possibly) pre-treated sample. After boiling for one minute and cooling under flowing water, the sample is made up to 100 mL with water and then measured at 525 nm in the photometer. A blank sample is treated in the same way.

### Interfering Factors
Interference caused by turbidity and larger amounts of organic substances is prevented as described above.

Chloride ions up to approx. 1000 mg/L are masked by addition of mercury ions. Larger chloride concentrations must be dealt with as described above.

## 6 Laboratory Measurements

Calculation of Results

Results are calculated by reference to the calibration chart.

### 6.2.1.13 Nitrate

Nitrate is found in many natural waters at concentrations between 1 and 10 mg/L. Higher concentrations often indicate the effects of nitrogen-containing fertilizers since the $NO_3^-$ ion is badly adsorbed in soil and so easily finds its way into the ground water. Very high nitrate concentrations are normally encountered in treated waste waters, a result of ammonium nitrogen being totally or partially oxidized to nitrate by microbiological action. On the other hand, ammonium dominates in raw waste waters. The parameter is of great importance in assessing the self-purification properties of water systems and the nutrient balance in surface waters and soil. The photometric method of determination with sodium salicylate is described below.

Areas of Application  ⟶   Water, waste water, soil

Apparatus

Spectrophotometer or fixed filter photometer (420 nm)
Water bath or sand bath
Porcelain crucibles (8 to 10 cm in diameter)

Reagents and Solutions

Sodium salicylate solution: 0.5 g sodium salicylate ($C_7H_5O_3Na$) is made up to 100 mL with water. The solution must be prepared freshly.

Alkaline tartrate solution: 400 g sodium hydroxide and 60 g sodium potassium tartrate ($C_4H_4O_6KNa \cdot 4\ H_2O$) is made up to 1 L with water.

Sulfuric acid: concentrated

Nitrate standard solution (1̶0̶0̶ mg/L, 10 mg/L): *1000 ppm*  1.6307 g potassium nitrate are made up to 1 L with water. 10 mL of this solution are then made up to 1 L.

Sample Preparation

The sample should be analyzed promptly, especially in the case of waste water. Strong discoloration can normally be removed by aluminum hydroxide precipitation (section 6.2.1.1).

Calibration and Measurement

The calibration is prepared by placing volumes of the standard solutions corresponding to 0.01 to 0.5 mg/L $NO_3^-$ into the porcelain crucibles. 2 mL of sodium salicylate solution is added and the mixture carefully evaporated to dryness on a water bath. The residue is dried for 2 hours

*0.01 to 0.5 mg $NO_3^-$ in Standards*
*(used 10.0 mL each of 0, 2, 4, 6, 8, 10 ppm $NO_3^-$)*

at approx. 100 °C. After cooling, 2 mL of concentrated sulfuric acid is given and left for 10 minutes. 15 mL water and 15 mL of the alkaline tartrate solution are then added. The solution is finally placed into a 100 mL graduated flask which is filled up with water and the photometric determination performed within 10 mins at 420 nm. The standard curve is constructed from the values thus obtained.   *10 mL of 10 ppm NO$_3^-$ gave 0.20 abs. units*

For determination in water, the sample volumes are chosen according to the expected NO$_3^-$ concentrations. In cases exceeding 100 mg/L, the samples must be diluted. Further analysis is carried out as for the standard curve. A blank sample is measured in the same way.

The determination in soil is carried out by extracting the soil with distilled water (1 : 10) and shaking for 2 hours. The extract is analyzed in the same way as water. In certain cases, the sample has to be filtered before measurement.

Interfering Factors
Interference caused by presence of more than 200 mg/L chloride can be prevented by dilution. Nitrite values exceeding 1 mg/L also interfere and may be eliminated by addition of 50 mg ammonium sulfate to the sample and evaporation to dryness before continuing with the analysis.

Calculation of Results
The amount of nitrate ions is determined by reference to the calibration curve and the blank value obtained.

## 6.2.1.14 Nitrite

Nitrite ions are found in unpolluted waters at levels not exceeding 1 μg/L. The ion can occur at greater concentrations as an unstable intermediate during nitrification of ammonium, especially where toxic effects are detected. It can be toxic to certain aquatic organisms at concentrations under 1 mg/L. Its presence is not permissible in drinking water.

Areas of Application  →  Water, waste water

Apparatus
Spectrophotometer or fixed filter photometer (540 nm)

Reagents and Solutions
Reagent solution: 0.5 g sulfanilamide ($C_6H_8N_2O_2S$) and 0.05 g N-(1-naphthyl)-ethylenediamine-dihydrochloride ($C_{12}H_{16}Cl_2N_2$) is dissolved in 25 mL water and 10.5 mL 36 % HCl. 13.6 g sodium acetate ($CH_3COONa \cdot 3\ H_2O$) is added and the solution made up to 50 mL. The solution is stable for several months.

| Nitrite standard solution (1000 mg/L, 1 mg/L): | 1.500 g sodium nitrite (dried for 1 hour at 105 °C) are made up to 1 L with water. The solution is stable for approx. one month at 4 °C. 1 mL of this solution is made up to 1 L with water. |
|---|---|

### Sample Preparation

The analysis should be carried out within a few hours of sample collection. In all cases, the sample must be kept cool until examination. Discoloration and colloidal turbidity may be removed by aluminum hydroxide flocculation (section 6.2.1.1). At pH $>10$ or base capacity exceeding 6 mmol/L, the pH value is adjusted to pH 6 with dilute hydrochloric acid.

### Calibration and Measurement

The standard curve is prepared as follows: 1 to 25 mL of the diluted nitrite standard solution are pipetted into 50 mL graduated flasks, diluted to approx. 40 mL and 2 mL reagent solution are added. The flasks are filled, the contents mixed and measured at 540 nm after leaving for 15 minutes.

Sample concentrations are measured using 40 mL (or a smaller volume made up to 40 mL). The pH should lie between 1.5 and 2. Further steps are as described above. A blank sample is measured in an identical manner.

### Interfering Factors

Nitrogen-oxides possibly present in the laboratory air can interfere with the determination. Strong oxidizing and reducing agents (active chlorine, sulfite) at high concentrations can also interfere.

### Calculation of Results

The nitrite concentration is calculated by reference to the calibration curve and blank value obtained.

### 6.2.1.15 Oils and Fats

In surface waters and waste waters or in ground water after spillages, contamination by mineral hydrocarbons and plant and animals fats can occur. Depending on their solubility, such substances can be present in solution, as emulsions or as the free phase. Emulsion formation is assisted by the presence of surfactants. The presence of oils and fats in raw water destined for drinking water is undesirable owing to taste and odor even at very small concentrations. Problems in waste water disposal can arise through the freeing of fatty acids which can lead to corrosion of concrete.

The gravimetric determination after extraction with trichlorotrifluoroethane is described below. Hexane or petroleum ether may be substituted for this solvent but their flammability should be considered. A differentiation between saponifiable fats (plant and animal) and unsaponifiable mineral components may be made.

## 6.2 Analytical Methods

Areas of Application ➡ Water, waste water

### Apparatus
Separating funnels, 1 L
Water bath
Centrifuge
Flasks, ground glass stoppers, 250 mL
Reflux condenser

### Reagents and Solutions
Trichlorotrifluoroethane ($C_2Cl_3F_3$) (or n-hexane or petroleum ether)
Ethanol
Alcoholic KOH, 0.1 M

### Sample Preparation
Samples should be collected in glass bottles, previously cleaned with solvents.

### Measurement
The extraction is performed in the sample bottle by shaking for approx. 1 minute after addition of 25 mL solvent. After separation, the aqueous phase is siphoned into a second bottle containing 25 mL of solvent and shaken again for 1 minute. The aqueous phase is then discarded. The extracts are passed into a 1 L separating funnel together with solvent washings from the bottles. The remaining water is shaken with the solvent in the funnel. The aqueous phase is allowed to run out (or solvent phase in the case of trichlorotrifluoroethane) and the solvent passed through a filter paper with a little anhydrous sodium sulfate into a weighed (constant weight) glass crucible (a platinum crucible is better). In the case of emulsions, centrifugation should be carried out. The solvent is removed over a water bath (max. 80 °C) the crucible dried for 5 minutes at 105 °C and allowed to cool in a desiccator before weighing the residue.

Should soaps be included in the analysis, the sample must be acidified to pH 1 to 2 so that the free fatty acids are extracted together with the oils and fats. Soaps may be determined separately by acidifying the pre-extracted sample and re-extracting.

The saponifiable oils and fats may be separated from the nonsaponifiable components as follows: the residue in the crucible is dissolved in ethanol, transferred to a 250 mL ground-glass stoppered flask and 50 mL of alcoholic KOH is added. After refluxing for 60 minutes, the content is transferred to a separating funnel together with 10 - 50 mL flask washings (solvent). The funnel is shaken, the phases separated and extraction of the alcoholic KOH continued by addition of more solvent. The combined extracts are then filtered into a crucible and treated according to the procedure described above. The residue after evaporation contains the non-saponifiable oils and fats.

### Interfering Factors

The determination is fairly unspecific as various other components such as emulsifying agents, waxes or surfactants can be wholly or partially extracted at the same time.

Emulsions can seriously impair the results but can be destroyed by addition of sodium sulfate or by acidification or centrifugation.

Strongly contaminated samples or sludges are firstly evaporated down on a water bath. Residues are quantitatively transferred to an extraction thimble and subjected to Soxhlet extraction for several hours. The extract is then treated as described above.

### Calculation of Results

Content of extractable oils and fats (mg/L) = $\dfrac{a \cdot 1000}{b}$

a = weight of extraction residue (mg)
b = sample volume (mL)

### 6.2.1.16 Phenol Index

Aromatic hydrocarbons with hydroxy groups attached to the aromatic ring are known as phenols. They are found at small concentrations in natural waters because phenolic compounds are components in plants. Additionally, they can be formed during humification processes occuring in soil. However, much higher concentrations appear in some industrial waste waters.

Phenols can be toxic towards water organisms and can accumulate in particular cellular components. Chlorination of phenol-containing water can lead to formation of chlorophenols with an unpleasant odor or taste.

The method described below includes the determination of total phenolic substances and the results are given in terms of phenol ($C_6H_5OH$). A distinction is made between the phenol index determined without distillation ("total phenols") and the determination with distillation ("steam-distillable phenols"). Analytically, the reaction of phenols and other reactable compounds with 4-amino-antipyrine to form antipyrine dyes is utilized.

Areas of Application ⟶ Water, waste water

### Sample Preparation

Immediately after collection, the pH value of the sample is adjusted to 12 with sodium hydroxide or to 3-4 with nitric acid.

a) Phenol Index without Distillation ("total phenols")

### Apparatus
Spectrophotometer or fixed filter photometer (460 nm or 510 nm)
Separating funnels

### Reagents and Solutions

Amino-antipyrine solution: 2 g 4-amino-2,3-dimethyl-1-phenyl-3-pyrazoline-5-one ($C_{11}H_{13}N_3O$) is dissolved in water and made up to 100 mL.

Buffer solution (pH 10): 34 g ammonium chloride and 200 g potassium sodium tartrate ($C_4H_4O_6KNa \cdot 4\,H_2O$) is dissolved in 700 mL water, treated with 150 mL ammonia solution (25 %) and made up to 1 L with water.

Peroxodisulfate solution: 0.65 g potassium peroxodisulfate ($K_2S_2O_8$) is dissolved in 1 L water.

Sodium sulfate, anhydrous
Copper sulfate ($CuSO_4\; 5\,H_2O$)
Chloroform ($CHCl_3$)

Phenol standard solutions (10 mg/L, 1 mg/L, 0.1 mg/L): 1 g phenol is made up to 1 L with water. The required standards are prepared by fresh dilution.

### Calibration and Measurement
Various standard solutions are prepared covering the concentration range of 0 to 100 µg/L and are treated in the same manner as the samples to be analyzed. 500 mL of sample, diluted if necessary, are adjusted to pH 4 and treated with 0.5 g copper sulfate. The solution is then transferred to a 1 L separating funnel and 20 mL buffer added. The pH is adjusted to 10 with sodium hydroxide if necessary. 3 mL amino-antipyrine solution are added followed by 3 mL peroxodisulfate solution after shaking for a short time. The mixture is further shaken and left in darkness for 30 to 60 minutes. The dye thus formed is extracted by thorough shaking for 5 minutes with 25 mL of chloroform. The chloroform phase is filtered through 5 g anhydrous sodium sulfate into a 25 mL volumetric flask which is then topped up with chloroform. The color intensity is measured in the photometer at 460 nm against chloroform as the reference. A blank sample consisting of 500 mL water is treated in an identical manner.

### b) Phenol Index after Distillation ("steam-distillable phenols")

### Apparatus
In addition to the equipment listed under a):
Distillation apparatus with 1 L round-bottom flask, Liebig condenser, heater, 500 mL measuring cylinder.

### Reagents and Solutions
In addition to the reagents listed under a):

Buffer solution (pH 4):     151 g disodium hydrogen phosphate ($Na_2HPO_4 \cdot 2 H_2O$) and 142 g citric acid ($C_6H_8O_7 \cdot H_2O$) are made up to 1 L with water.

Ferric sulfate ($Fe_2(SO_4)_3 \cdot 9 H_2O$)

Potassium hexacyano-ferrate (III) solution:     8 g ferricyanide ($K_3Fe(CN)_6$ is made up to 100 mL with water. The solution must be stored away from light and is stable for approx. 1 week.

### Calibration and Measurement
500 mL of sample (possibly diluted) is placed in the distillation flask and 0.5 g copper sulfate is added. The solution is shaken several times over a period of 10 minutes and 50 mL buffer solution (pH 4) is added. If necessary, the pH value is adjusted to 4. 400 mL are then distilled over into the measuring cylinder which is topped up to 500 mL with water and the contents transferred together with 20 mL buffer solution (pH 10) to the separating funnel. 3 mL amino-antipyrine solution is added and, after shaking for a short time, followed by 3 mL potassium ferricyanide solution after which the shaking is continued. The further procedure is as described under a). At higher concentrations of phenolic substances, the chloroform extraction step can be dispensed with. In such cases, a smaller sample volume e.g. 100 mL is employed and the photometric measurement performed at 510 nm.

### Interfering Factors
For a) and b)
Oxidizing agents present in the sample (e.g. chlorine or chlorine dioxide) interfere but their effects can be minimized by addition of ascorbic acid. In the presence of reducing agents, method b) is employed, a small amount of ferric sulfate being added before the distillation. Should the sample have its own color, a blank without addition of amino-antipyrine is treated similarly and its value substracted from the result.

### Calculation of Resuluts
The phenol concentrations are obtained from the calibration curves by reference to the blank.

### 6.2.1.17 Phosphorus Compounds

Phosphorus compounds can be determined in water, waste water or soil in various forms: total phosphorus, o-phosphate, hydrolyzable phosphate and organically bound phosphate. The method of analysis depends on the forms present and the aims of the examination.

Natural unaffected waters mostly contain total phosphorus compounds at concentrations of less than 0.1 mg/L. Phosphorus compounds are fixed in the soil to such an extent that the danger of seepage to the deeper layers or even into the groundwater is relatively slight. However, surface waters can have significant phosphorus concentrations caused by soil erosion or effluent introduction. Although inorganic phosphorus compounds are not toxic, they are undesirable components of rivers and lakes used as sources of raw water for drinking water preparation owing to the danger of eutrophication.

Phosphates which are to be determined in water samples without prior hydrolysis are known as "orthophosphate". Condensed phosphates are converted into orthophosphates by hydrolysis of the sample under weakly acidic conditions. During this process, conversion of certain organic phosphate components into inorganic phosphates is unavoidable. Only that portion of the organic P-compounds which reacts under strong oxidizing conditions is referred to as "organic phosphorus". Problems are also encountered in distinguishing between dissolved and suspended phosphorus, as conventionally the type of filtration (0.45 µm) determines the phosphorus species measured.

For problems concerning the eutrophication of water systems, the determination of dissolved o-phosphates and the total phosphate is recommended. At higher concentrations of both parameters, conditions favourable for eutrophication are always found whereas at higher total phosphate and non-detectable amounts of o-phosphate, P-limited growth conditions are often indicated. The determination of the sum of o-phosphates and condensed phosphates is, however, often necessary in the cases of drinking water and water for industrial use (dosing with phosphates to reduce corrosion) as well as in waste water.

Areas of Application ⟶ Water, waste water, soil

## a) o-Phosphate

### Apparatus
Spectrophotometer or fixed filter photometer (880 or 700 nm)

### Reagents and Solutions

Ammonium molybdate solution: 40 g ammonium molybdate $((NH_4)_6Mo_7O_{24} \cdot 4 H_2O)$ is made up to 1 L with water

Ascorbic acid solution: 2.6 g L(+) ascorbic acid $(C_6H_8O_6)$ is dissolved in 150 mL water. This solution must be freshly prepared.

Potassium antimony (III)oxytartrate solution: 2.7 g potassium antimony(III)oxytartrate $(K(SbO)C_4H_4O_6 \cdot \frac{1}{2} H_2O)$ is dissolved in water and made up to 1 L.

Reaction mixture: 250 mL of 25% sulfuric acid, 75 mL ammonium molybdate solution and 150 mL ascorbic acid are mixed. 25 mL potassium antimony(III)-oxytartrate solution are added and the solution

| | |
|---|---|
| | mixed. The solution should be freshly prepared and stored in the refrigerator. |
| Phosphate standard solution (1 mg/L $PO_4^{3-}$): | 139.8 mg potassium dihydrogenphosphate ($KH_2PO_4$) dried at 105 °C, are made up to 1 L with water. 10 mL of this solution are taken and made up to 1 L. |

### Sample Preparation

The sample is passed through a 0.45 μm membrane filter as soon as possible after collection. The first 10 mL are discarded.

### Calibration and Measurement

Volumes of the standard phosphate solution corresponding to between 0 and 0.04 mg $PO_4^{3-}$ are placed in a series of 50 mL volumetric flasks and these made up to 40 mL with water. These solutions and max. 40 mL of a sample are then treated with 8 mL reaction mixture and made up to 50 mL with water. After mixing, the solutions are left for 10 minutes and measured at 880 or 700 nm in the photometer.

The blank correction for color and turbidity is made by adding to the water sample 8 mL of a reaction mixture containing only sulfuric acid and potassium antimony(III)oxytartrate in the amounts given above.

### Interfering Factors

Silicic acid at a concentration exceeding 5 mg/L can appear as a high phosphate concentration. Chromate has the opposite effect and can be reduced by adding 1 mL ascorbic acid solution. Sulfides exceeding 2 mg/L can be removed by addition of several mg of potassium permanganate. After shaking, the excess reagent is reduced by addition of 1 mL ascorbic acid solution.

### Calculation of Results

The phosphate concentration is obtained from the calibration curve
Conversion:                1 mg $PO_4^{3-}$ ≙ 0.326 mg P

## b) Hydrolyzable Phosphate (total inorganic phosphate)

### Reagents and Solutions

In addition to the solutions listed under a):
Conc. sulfuric acid
Sodium hydroxide, 20 %

| | |
|---|---|
| Phenolphthalein solution: | 1 g phenolphthalein is dissolved in 100 mL ethanol and treated with 100 mL water. |

### Measurement

Up to 40 mL of the sample are treated with 1 mL conc. sulfuric acid. After boiling for 5 minutes, cooling and neutralization using phenolphthalein solution and sodium hydroxide, the mixture is diluted to approx. 40 mL. The further procedure is as described under a). A blank is treated in the same way as the sample.

### Calculation of Results

The hydrolyzable phosphate is obtained from a calibration curve. The result shows the sum of o-phosphate and condensed phosphate. The hydrolyzable phosphate is obtained by subtracting the o-phosphate from this value.

#### c) Total Phosphate

### Reagents and Solutions

In addition to those solutions described under a) and b):

Potassium peroxodisulfate solution: 5 g potassium peroxodisulfate ($K_2S_2O_8$) is dissolved in 100 mL water. The solution must be freshly prepared.

### Measurement

100 mL sample (or a smaller volume, diluted to 100 mL) are treated with 0.5 mL conc. sulfuric acid so that a pH value of $<1$ results. 15 mL potassium peroxodisulfate solution are added and the mixture is boiled gently for 30 minutes (90 minutes in the case of resistant organic phosphorus compounds). After cooling, a drop of phenolphthalein solution is added together with sufficient sodium hydroxide solution to give a pink color. The solution is made up to 100 mL with water and the phosphate determined as described under a). The standard solution and the blank are treated in the same way.

### Calculation of Results

The concentration of total phosphorus is determined by reference to the calibration curve. The concentration of organically-bound phosphorus is obtained by subtracting the amount of hydrolyzable phosphate from the total phosphate.

### 6.2.1.18 Potassium

The concentration of potassium in natural waters seldom exceeds 20 mg/L whereas in some waste waters and especially seepage from waste dumps very high concentrations can be found, even exceeding the sodium level.

Areas of Application ⟶ Water, waste water, soil

### Apparatus

Flame photometer with 768 nm filter

## 6 Laboratory Measurements

### Reagents and Solutions

Potassium standard solutions (100 mg/L, 10 mg/L): 1.907 g potassium chloride (dried at 105 °C) are made up to 1 L with water. The solution contains 1000 mg/L potassium. Standard solutions containing 100 mg/L and 10 mg/L are prepared by dilution. The solutions are stored in plastic bottles.

### Sample Preparation
Water samples are filtered before measurement to prevent blockage of the suction equipment.

For determination of the available potassium in soil, 1 part of air-dried soil is shaken with 10 parts of distilled water and the filtered extract is analyzed.

### Calibration and Measurement
A standard curve with the emission intensity at 768 nm is constructed using standards in the desired concentration range (e.g. 0 to 1 mg/L, 0 to 10 mg/L). The sample and blank are then measured. Alternatively, the standard addition technique may be employed to compensate for possible matrix effects. For this purpose, the sample of unknown potassium content is treated with various known amounts of potassium. The measured intensities are plotted against the added amounts and the curve extrapolated. The point of intersection then gives the unknown potassium concentration (Figure 42).

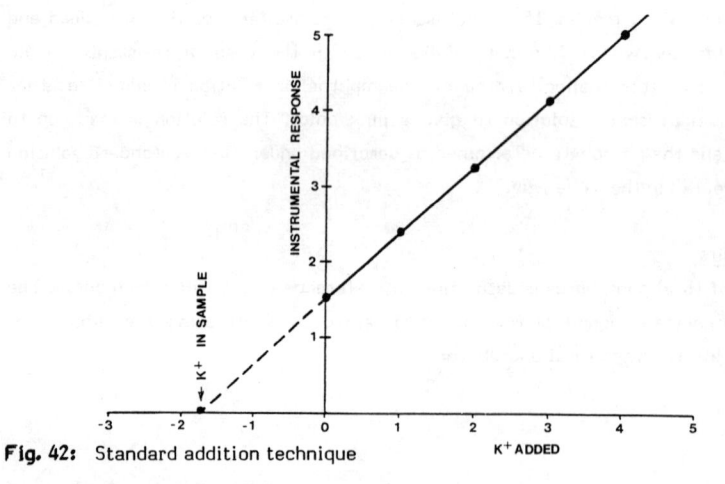

**Fig. 42:** Standard addition technique

Comparison of this curve with that obtained using pure standard solutions gives immediate information concerning interference or matrix effects.

### Interfering Factors
The presence of sulfate, chloride or bicarbonate at high concentrations can interfere. The standard addition method can almost entirely prevent such problems.

## Calculation of Results
Results are obtained by reference to the calibration curve.

### 6.2.1.19 Silicic Acid

Silicon occurs as one of the most abundant elements in all rocks and sediments. Silicon compounds such as silicic acid can be dissolved from such materials by weathering processes and so reach the water cycle. Silicic acid can be found in dissolved, colloidal or suspended form. The concentration in natural waters lies usually in the range from 0 to 20 mg/L but higher concentrations can be found in some strongly mineralized waters. Silicic acid is an undesirable component in waters for industrial use as deposits can be formed in pipes or boilers.

The method described here is the determination of the reactive "soluble" silicic acid after reaction with molybdate.

## Area of Application → Water

## Apparatus
Spectrophotometer or filter photometer (812 or 650 nm)

## Reagents and Solutions

Ammonium molybdate solution: 10 g ammonium molybdate $((NH_4)_6Mo_7O_{24} \cdot 4\,H_2O)$ is made up to 100 mL with warm water. After filtration and adjustment of pH to 7, the solution is stored in a plastic bottle.

Reducing solution: 0.5 g of 1-amino-2-naphthol-4-sulfonic acid $(C_{10}H_9O_4NS)$ and 1 g sodium sulfite (anhydrous) are dissolved in 50 mL water. The solution is then added to a solution containing 30 g of sodium hydrogen sulfite in 150 mL water. The reducing solution is stored in a plastic bottle in a refrigerator.

Oxalic acid solution: 10 g of oxalic acid $(C_2H_2O_4 \cdot 2\,H_2O)$ is made up to 100 mL with water.

Hydrochloric acid 20 %

Sodium carbonate solution: 25 g of anhydrous sodium carbonate is dissolved in 1 L water.

Silicic acid standard solution (10 mg/L $SiO_2$): A ready-made standard solution in ampoules is used. Where these are not available, standards should be prepared as follows:

1 g of silicon dioxide is heated in a platinum crucible for approx. 1 hour at 1100 °C. After cooling, 5 g of anhydrous sodium carbonate is added. The mixture is heated until it melts. Splashes must be avoided. After cessation of gas development, heating is continued at light red hot temperature for approx. 10 minutes. On cooling, the melt is dissolved in water and made up to 1 L. The solution is stored in a plastic bottle. It contains 1000 mg/L $SiO_2$. 10 mL of this solution are taken and made up to 1 L.

### Sample Preparation

Before determination of the dissolved silicic acid, the sample is passed through a 0.45 µm membrane filter. If the colloidal silicic acid is also to be determined, the sample must be dissociated: 100 mL of sample are placed in a platinum crucible together with 20 mL sodium carbonate solution and carefully evaporated down to 80 mL. This solution is transferred to a 100 mL volumetric flask. 5 mL hydrochloric acid (20 %) are added and the flask topped up.

### Calibration and Measurement

1 to 10 mL aliquots of the silicic acid standard solution are taken and made up to 50 mL. These solutions are measured together with 50 mL of sample by adding 1 mL of hydrochloric acid (20 %) and 2 mL of ammonium molybdate. After mixing, 1.5 mL of oxalic acid solution are added and mixing is continued. After 5 minutes, measurement is carried out in the photometer at 812 or 650 nm.

Owing to the lower sensitivity at 650 nm, 0 to 30 mL of standard should be used to construct the calibration curve. A blank is treated in the same way as the sample.

### Interfering Factors

Phosphates, iron and sulfides can interfere. Interference by phosphate is reduced by addition of oxalic acid. Color interference can be compensated by photometric comparison measurements.

### Calculation of Results

The content of silicic acid is obtained by reference to the calibration curve.

### 6.2.1.20 Sodium

Sodium is one of the major components found in natural waters. Very small concentrations are only encountered in rain water. Very high concentrations are found in brines and sea water. In soils, especially of arid regions, the content of sodium ions plays an important role in the problem of oversalting.

Areas of Application ⟶ Water, waste water, soil

### Apparatus

Flame photometer with 589 nm filter

### Reagents and Solutions

Sodium standard solutions (100 mg/L, 10 mg/L):  2.542 g sodium chloride (dried at 105 °C) is made up to 1 L with water. This solution contains 1000 mg/L of sodium. Standard solutions containing 100 mg/L and 10 mg/L are prepared by dilution. The solution are stored in plastic bottles.

### Sample Preparation

Water samples are filtered before analysis in order to prevent blockage of the dosing apparatus.

For determination of the available sodium in soil, one part of air-dried soil is shaken with 10 parts of distilled water and the filtered extract is analyzed.

### Calibration and Measurement

A standard curve is constructed by measuring the emission intensity (589 nm) of standards in the required working range (e.g. 0 to 1 mg/L, 0 to 10 mg/L). The sample and blank are then measured. Alternatively, the standard addition technique may be employed to compensate for possible matrix effects. For this purpose, various known amounts of sodium are added to the sample of unknown concentration. The measured intensities are plotted against the amounts added and the curve extrapolated. The point of intersection then gives the sodium concentration in the sample (Figure 42).

Comparison of this curve with that obtained using pure standard solutions gives immediate information concerning possible matrix effects or interferences.

### Interfering Factors

The presence of sulfate, chloride or bicarbonate at concentrations exceeding 1000 mg/L can interfere. The standard addition method can almost entirely eliminate such problems.

### Calculation of Results

Results are obtained by reference to the standard curve.

### 6.2.1.21 Sulfate

Sulfate ions occur in natural waters at concentrations up to 50 mg/L. Concentrations of up to 1000 mg/L can be found in waters having contact with certain geological formations e.g. gypsum reserves, water from pyrite quarries.

Contaminated waters and waste water normally have high sulfate concentrations so that increased sulfate drinking water normally points to the influence of waste water. Two methods for sulfate determination are described below: the gravimetric method, employed where accuracy is important and the turbidimetric method, less accurate but quicker to carry out.

Areas of Application ⟶  Water, waste water, soil

a) Gravimetric method

Apparatus
Laboratory oven
Muffel furnace
Water bath
Quartz or platinum crucibles

Reagents and Solutions

Barium chloride solution:  10 g barium chloride ($BaCl_2 \cdot 2\ H_2O$) is dissolved in 90 mL of water.

Methyl orange indicator:  100 mg methyl orange is made up to 100 mL with water.

Silver nitrate solution:  1 g silver nitrate is dissolved together with a few drops of nitric acid in 100 mL water.

Hydrochloric acid, 20 %
Sodium chloride solution, 10 %

Sample Preparation
Suspended particles must be filtered off before the determination. Silicates interfere at concentrations in excess of 25 mg/L and can be removed together with organic substances as follows:

An aliquot of sample containing no more than 50 mg/L sulfate is evaporated almost to dryness on the water bath. A few drops of 20 % hydrochloric acid and 10 % sodium chloride are added and the sample evaporated to dryness so that the salt crust is in contact with the acid. After ashing at approx. 500 °C, the residue is moistened with 3 mL water and a few drops of hydrochloric and again evaporated to dryness. The sample is then dissolved in a little hot water and 1 mL hydrochloric acid. Approx. 50 mL of hot water are added and the solution filtered while still hot. The filter residue contains the insoluble silicic acid and is washed with water until no chloride is indicated by testing with silver nitrate solution. The filtrate and washings are used for the sulfate determination.

Measurement
A volume of sample containing up to 50 mg/L sulfate is placed into a beaker and made up to 200 mL, if necessary. The solution is adjusted to pH 7 with the aid of the methyl orange indicator, 2 mL 20 % hydrochloric acid is added and the mixture boiled for a short time. Hot

barium chloride solution is then added with stirring until the precipitation appears to be complete and a further 3 mL is added. The heating is continued for a further 1/2 hour and the mixture left for at least 2 hours (best overnight) before filtering through ashfree filter paper or a constant-weight porcelain sintered disk (A 1). The precipitate is washed with hot water until the chloride reaction is negative. The filter paper is transferred to a constant-weight porcelain crucible and carefully dried and ashed during which the paper should not be permitted to burn in the open flame before heating for 30 minutes at approx. 800 °C. In the case of a filter cup, it should be heated to constant weight at 300 °C and then weighed after cooling in a desiccator.

Interfering Factors

Interference by organic compounds, nitrates and silicic acid can be prevented by the sample preparation as described. Heavy metals can cause lower values to be obtained as they hinder the complete precipitation of barium sulfate.

Calculation of Results

$$\text{Sulfate (mg/L)} = \frac{a \cdot 411.5}{b}$$

a = weight of barium sulfate (mg)
b = sample volume (mL)

b) Turbidimetric method

Apparatus

Magnetic stirrer
Spectrophotometer or fixed filter photometer (420 nm)

Reagents and Solutions

Barium chloride ($BaCl_2 \cdot 2 H_2O$), crystalline

Conditioning reagent: 30 mL conc. hydrochloric acid, 300 mL water, 100 mL ethanol (approx. 95%) and 75 g sodium chloride are mixed. 50 mL glycerine ($C_3H_8O_3$) is added and the solution mixed.

Standard sulfate solution (100 mg/L $SO_4^{2-}$): 147.9 mg anhydrous sodium sulfate are made up to 1 L with water.

Sample Preparation

Cloudy samples are filtered before analysis. Samples having concentrations in excess of 50 mg/L must be diluted.

Calibration and Measurement

Standard solutions with sulfate concentrations in the range 0 to 50 mg/L and the sample are made up to volumes of 100 mL, if necessary. 5 mL of conditioning reagent is added and the solutions are continuously stirred. 0.2 to 0.3 mg solid barium chloride is added during stirring and the stirring continued for exactly one minute. Immediately afterwards, the solution is transferred to a cuvette and the absorption at 420 nm measured repeatedly over a period of 2 to 3 minutes. The highest measured value is noted. A blank, consisting of a water sample without addition of barium chloride is treated in the same way.

Calculation of Results
The concentration of sulfate ions is obtained by reference to the calibration curve.

6.2.1.22 Surfactants

Surfactants are mainly synthetic surface active substances which have hydrophilic and hydrophobic properties and are extensively used in the home and in industry. Anionic surfactants today form the largest part of the total production but the proportion of non-ionic surfactants is increasing.

Surfactants reach water systems via waste water pipes and can lead to problems (reduction of oxygen diffusion, foaming). In many countries, the application of the so-called "hard" surfactants i.e. not easily biologically degradable materials instead of long-chain easily degradable surfactants is no longer permitted.

A method for determination of anionic surfactants by the methylene blue reaction and also a rapid test for determination of non-ionic surfactants with the Dragendorff reagent is described below.

Areas of Application ⟶ Water, waste water

a) Determination of anionic surfactants

Apparatus
Spectrophotometer or fixed filter photometer (650 nm)
Separating funnel, 500 mL

Reagents and Solutions
Methylene blue solution: 30 mg of methylene blue are dissolved in 500 mL of water and treated with 6.8 mL conc. sulfuric acid and 50 g of sodium dihydrogenphosphate ($NaH_2PO_4 \cdot H_2O$). The mixture is then made up to 1 L with water.
Chloroform
Wash solution: 6.8 mL conc. sulfuric acid are dissolved in water together with 50 g of sodium dihydrogenphosphate ($NaH_2PO_4 \cdot H_2O$) and made up to 1 L.

Sodium hydroxide, 1 M

Surfactant standard solutions (1000 mg/L, 10 mg/L): 1 g sodium lauryl sulfate is made up to 1 L with water. The solution is stored in a refrigerator. 10 mL of this solution is made up to 1 L with water (fresh).

### Sample Preparation

The sample is to be diluted according to the expected amount of methylene blue active substance (MBAS). At concentrations of 10 to 100 mg/L MBAS, 2 mL of sample is used; at 2 to 10 mg/L MBAS, 20 mL is used and at smaller concentrations, 100 mL to 400 mL. Before measurement, the samples are made weakly alkaline by dropwise addition of 1 M sodium hydroxide using phenolphthalein as indicator. The weak pink color obtained is then just removed by adding dilute sulfuric acid.

### Calibration and Measurement

For construction of the calibration curve, various volumes of the dilute standard solution between 0 mL and 20 mL are placed into 5 separating funnels. The standard solutions and the sample itself are diluted to 100 mL and shaken for 30 seconds after addition of 10 mL chloroform and 15 mL methylene blue solution. The chloroform extract is transferred to a second separating funnel after phase separation and extracted twice with 8 mL chloroform. The combined extracts are shaken for 30 seconds with 50 mL wash solution and passed through glass wool into a 50 mL volumetric flask. The wash solution is extracted twice with 10 mL each of chloroform and the extracts also passed into the volumetric flask. The glass wool is then rinsed with some chloroform into the flask which is then filled with chloroform to 50 mL. The extinction is measured at 650 nm within one hour.

A blank consisting of 100 mL water is treated in the same way as the sample.

### Interfering Factors

The detection of surfactants can be affected by the presence of certain substances lacking surfactant properties e.g. aromatic sulfonates, organic phosphates.

Interference through proteins or alkaline salts of higher fatty acids is minimized by the buffering.

Sulfide or thiosulfate interference can be prevented by addition of a few drops of hydrogen peroxide. Chloride concentrations exceeding 1000 mg/L interfere so that such samples should be diluted to lower concentrations before the determination.

### Calculation of Results

The results are obtained by reference to the calibration chart obtained.

### b) Determination of Non-ionic Surfactants

The screening method described here for estimation of non-ionic surfactants has a detection limit of 0.1 mg/L. For more exact measurements of smaller concentrations or elimination of matrix effects, a pre-concentration (clean-up) is necessary. Such techniques may be obtained by reference to specialist hand-books (Literature list: Deutsche Einheitsverfahren; US-Standard Methods).

Reagents and Solutions

Bismuth salt solution: 1.7 g basic bismuth nitrate ($BiO_xNO_3 \cdot H_2O$) is dissolved in 20 mL glacial acetic acid and made up to 100 mL with water.

Potassium Iodide solution: 40 g potassium iodide is dissolved in 100 mL water.

Barium Chloride solution: 20 g barium chloride ($BaCl_2 \cdot 2 H_2O$) is made up to 100 mL with water.

Reagent (ready to use): The bismuth and potassium iodide solutions are combined, treated with 200 mL glacial acetic acid and made up to 1 L with water. 100 mL of this solution are treated with 50 mL barium chloride solution. The reagent is stable for approx. 14 days when stored in brown bottles.

Measurement

The water sample to be measured is filtered. 5 mL of filtrate are treated with 5 mL of reagent in a test tube and shaken. The presence of non-ionic surfactants of the polyalkylene oxide type is indicated by an orange-red precipitate. Very small amounts are indicated by a cloudiness. After centrifugation, the precipitate is easily recognizable at the bottom of the tube.

The concentration of non-ionic surfactants can be semi-quantitatively estimated by visual comparison with standard containing nonylphenol ($C_{15}H_{24}O$) at concentrations of 0.1 to 5 mg/L.

6.2.1.23 Zinc

Zinc is present in natural waters in the range up to 50 µg/L. Increased concentrations in the drinking water range are generally a result of corrosion of galvanized steel pipes so that, after a long standing time, values of up to 5 mg/L are not unusual. Zinc can be found at levels up to several mg/L in surface waters. For some fish species, the toxic threshold can lie below 0.5 mg/L.

Zinc is determined using the complexing reaction with dithizone.

Area of Application ⟶ Water

### Apparatus
Spectrophotometer or fixed filter photometer (535 nm)
Separating funnel (100 mL)

### Reagents and Solutions

Dithizone solution: 10 mg dithizone ($C_{13}H_{12}N_4S$) is dissolved in 1 L chloroform. The solution is stored in a brown glass bottle.

Acetate buffer solution:
a) 60 g sodium acetate is made up to 1 L with water.
b) 125 mL glacial acetic acid is made up to 1 L with water.
1 part of solution a) is mixed with 1 part of solution b) to obtain the buffer solution.

Sodium thiosulfate solution, 25 %

Zinc standard solution (10 mg/L Zn): 4.399 g zinc sulfate is made up to 1 L with water. 10 mL of this solution is made up to 1 L.

### Sample Preparation
If the solutions are clear, a sample preparation is not necessary.

### Calibration and Measurement
0 to 10 mL of the zinc standard solution are taken (corresponding to 0 to 0.1 mg Zn) and treated in the same way as the sample.

The sample pH is adjusted to 2-3 with hydrochloric acid and 10 mL of this solution transferred to the separating funnel. 5 mL of acetate buffer and 1 mL of sodium thiosulfate solution are added and the pH checked to be between 4 and 5.5. Then, 10 mL dithizone solution are added and shaken for 3 minutes before filtering the organic phase. The photometric measurement is carried out at 535 nm.

### Interfering Factors
In addition to zinc, other elements such as silver, copper, nickel, cadmium and lead form colored complexes with dithizone. These elements are almost completely masked by the addition of thiosulfate solution.

### Calculation of Results
The zinc content is obtained by reference to the calibration curve.

## 6.2.2 Microbiological Testing of Water

As water can carry a number of different pathogenic organisms to a large number of consumers and over a wide area, the early recognition of the relevant contamination must be ensured.

This monitoring of drinking, industrial, bathing and other waters is carried out by microbiological water tests. In general, the tests involve deter-mination of the total count of virulent organisms and identification of special organisms indicative of hygienically suspect contamination (e.g. Escherichia coli and coliform bacteria) or even pathogens themselves. Of the pathogens and facultative pathogenic types which can occur in water, the bacteria of the family Enterobacteriaceae are of particular importance. The species Salmonella, Shigella and Escherichia belong to this as well as the so-called "coliform bacteria" and Proteus, Yersinia and Erwinia. Salmonella and Shigella are classed as being particularly pathogenic, the others being classed as facultatively pathogenic.

In hygienic water testing, emphasis is mainly placed on the organisms mentioned above but further hygienically important bacteria can be present such as Vibrio cholerae (cause of cholera), mycobacterium tuberculosis (tuberculosis), Clostridium tetani (tetanus) or Bacillus anthracis (anthrax). In addition, the eggs of various parasites can be present in water.

### 6.2.2.1 Sampling, Transport and Storage of Water Samples for Microbiological-hygienic Tests

Sampling is carried out using sterile glass stoppered bottles where the stopper and neck are covered with aluminum foil to prevent contamination. Before sterilization, sodium thiosulfate solution (1 M) should be placed in the bottles in order to bind any chlorine or chloramine present. 0.1 mL of this solution is placed in a 100 mL bottle, 0.25 mL in a 250 mL bottle and 0.5 mL in a 500 mL bottle. Separate bottles should always be used for the microbiological tests and should only be 5/6 filled in order to facilitate the shaking necessary before examination.

Before sampling from taps, the tap should be fully opened and closed several times to get rid of dirt particles. The tap exit is then flamed for a sufficiently long time that a quenching noise is heard on opening. The water is then allowed to run free in a pencil-thick stream for approx. 5 minutes before filling the bottle, closing it under sterile conditions and labelling correctly.

In the case of a well with hand-pump, the exit is flamed until completely dry. The well is then uniformly pumped for about 10 minutes during which time attention must be paid to making sure that the pumped water does not run back into the well or be allowed to seep near the well.

In the case of a container or reservoir, the samples are collected in bottles about 30 cm under the surface with special holders attached to poles of various lengths.

At high environmental temperatures, the samples must be cooled in order to prevent increases in the count after sampling. All samples must be protected against breakage during transport to the laboratory.

The samples should be examined immediately on arrival in the laboratory. Should this not be possible, they must be stored at approx. +4 °C. Under no circumstances should the period between sampling and testing exceed 48 hours (even on cooling). Where this is not possible, the microbiological tests must be performed on-site e.g. in a mobile laboratory.

#### 6.2.2.2 Technical Requirements for the Hygienic Water Examination

##### The Working Area

Micro-organisms are present everywhere so that the microbiologist must protect his working area and samples from the danger of contamination.

Rooms can be disinfected by the action of UV light. The working surfaces should be smooth and easy to clean and disinfect e.g. stainless steel.

##### Cleaning and Sterilization

Sterilization of the apparatus employed in microbiological testing of water is an essential requirement.

Apparatus must only be sterilized by heating. Culture dishes and other plastic equipment are supplied in pre-packed sterile containers.

The following equipment is necessary for sterilization:

Laboratory oven with or without air circulation
Steam generator
Autoclave

New glass apparatus is mechanically cleaned for a short time with mains water and rinsed with acidified and then distilled water.

Used glass equipment which was employed for examination of impeccable drinking water should be cleaned mechanically with alkali, placed in acidified water, rinsed with distilled water and finally dried. The various pieces of equipment employed in the testing of contaminated water e.g. glass culture dishes with medium should first of all be sterilized in an autoclave at 120 °C for 15 minutes and then cleaned as above. Disposable articles are boiled or autoclaved for 1/2 hour with water and disinfectant before disposal. Glass equipment which has been cleaned and plugged with cotton wool is sterilized for 2 hours at 160 °C in a laboratory oven. Before sterilization of glass apparatus incorporating ground glass joints, a paper strip (about 1 cm broad) is placed between the neck and stopper which can subsequently be removed. The stopper and neck are then covered with aluminum foil.

Pipettes are sterilized in pipette tins. The air holes are kept open during sterilization but are closed afterwards.

Glass petri dishes, reagent bottles and Erlenmeyer flasks are sterilized in wire mesh baskets.

With the exception of the petri dishes, all apparatus sterilized as described above remain in a sterile condition for a considerable period of time. Sterilization should be repeated after six weeks of storage.

Plastic culture dishes delivered in sealed sterile plastic bags remain sterile in an unopened condition for more than a year.

Heat resistant culture media (e.g. agar-agar) are best sterilized by pressurized steam in an autoclave at approx. 120 °C ($\triangleq$ 1 bar pressure) for 20-30 minutes.

Thermolabile culture media (e.g. gelatine) are fractionally sterilized i.e. the media are left in circulating steam for 30 minutes on 3 consecutive days and incubated in the interim periods at + 25 °C. In this way, even heat resistant bacteria and/or fungal spores are destroyed without affecting the growth properties of the medium.

### 6.2.2.3 Carrying out Microbiological Water Tests

At the present time, 2 procedures are basically followed:

a) determination of the total count in 1 mL of water sample
b) detection of Escherichia coli and coliform bacteria in a definite water quantity (mostly 100 mL).

### 6.2.2.3.1 Total Bacterial Count

The total bacterial count is the number of colonies visible under a magnification of 6-8 which have developed under defined conditions. It provides a measure of the degree of microbiological contamination of the water and especially of sudden bacterial invasions.

One mL of well mixed water sample is pipetted into a sterile petri dish and mixed with 10 mL of sterile nutrient gelatine or agar. Where higher counts are expected, diluted cultures should be prepared with sterile water (e.g. 1:100, 1:1000 etc). The gelatine should be stored under sterile conditions in tubes, liquified at +35 °C in the water bath and cooled to +30 °C before pouring. Tubes of nutrient agar are liquefied in boiling water and cooled to +46 °C $\pm$ 2 °C before pouring. The tube openings must be flamed before pouring into the petri dishes.

The petri dishes are swung in a figure 8 movement to achieve thorough mixing and then left horizontally until the contents solidify. Cooling may be necessary as nutrient gelatine solidifies at temperatures below + 25 °C.

The culture plates are then inverted and incubated at +20 °C and/or 37 °C for 44 h $\pm$ 4 h.

The visible colonies are then counted with the aid of a 6-8 x magnifying glass. Using agar and an incubation temperature of +37 °C, a first count can be performed after 20 h $\pm$ 4 h.

The use of a counting plate (e.g. according to Wolfhügel) or another suitable counting aid is recommended for heavily populated plates.

The total bacterial count is given in terms of 1 mL of water sample. A value exceeding 100 is rounded up to the nearest ten; a value exceeding 1000 is rounded up to the nearest hundred, etc.

In addition, the type of medium and the incubation time and temperature must be given.

Example:
Total bacterial count (gelatine, 44 h, + 20 °C): 90 colonies/mL.

Further methods (but not official) include:
Total count determination by inoculation of agar plates:
Sterile agar plates having a dry surface are inoculated with a known sample volume evenly distributed over the surface with a special spatula (e.g. Drigalski).

Dropping method according to Miles and Misra:
Five drops of sample are allowed to fall approx. 20 mm from a calibrated pipette on to a well-dried agar plate. The total count per mL is then calculated from the mean count of micro-colonies per plate.

Membrane filter method:
Culture growth on membrane filters is possible because the microorganisms cannot penetrate very deeply. If a filter retaining microorganisms is placed on plates of solid medium (agar or filter plates soaked in nutrient solution), the nutrients can diffuse through the filter allowing colonies to be formed on the filter surface.

The advantage of the membrane filter method is that even at very low microorganism concentrations requiring large sample volumes, determination of total count is possible.

Also, any growth inhibitors present in the sample are removed by the filtration process and cannot affect the subsequent growth of microorganisms.

Dip slide method
A sterile glass plate having the dimensions of a normal microscope slide (26 x 76 mm) is fixed

underneath the cap of a suitable cylindrical and sterile container (Fig. 43). The slide is coated on one or both sides with a sterile agar medium. Such slides are available commercially.

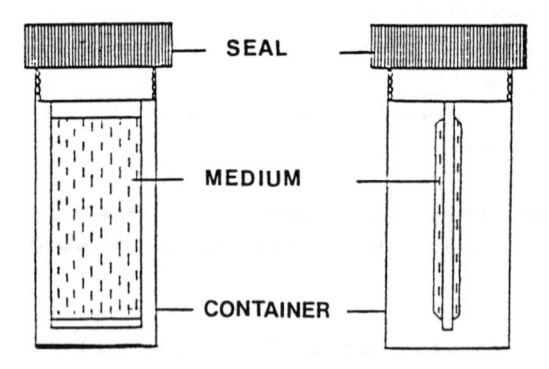

**Fig. 43:** Apparatus for determination of total bacterial count by the dip slide method

The inoculation is carried out by briefly dipping the glass slide into the water sample or a dilution thereof. Excess sample is drained off the bottom of the slide by briefly blotting on filter paper.

Approximately the same sample volume always remains on the surface of the agar medium. The glass slide is then placed in the container for incubation, after which the bacterial count is obtained by comparing the colony density with standard slides. The colonies may also be counted although there is no advantage in this because of the low accuracy of the dip slide method. It is suitable for total counts of about $10^3$ per mL.

### 6.2.2.3.2 Escherichia Coli and Coliform Bacteria

Liquid Enrichment Procedure

As long as the only question is whether Escherichia coli and/or coliform bacteria are present in 100 mL water, it suffices to mix 100 mL of sample with 100 mL of a doubly concentrated lactose-peptone solution. After incubation at +37 °C for $20 \pm 4$ hours, the mixture is examined for production of acid and gas. If no acid and gas are produced the water is satisfactory (as regards E. coli and coliform bacteria) for drinking water requirements and the test procedure can be stopped. On detection of lactose fermentation with gas and acid production, it must be determined whether the organisms reponsible are E.coli, coliform bacteria or bacteria of a different group.

For this purpose, a small quantity of the turbid fermented lactose-peptone solution is taken on a sterile platinum loop and fractionally streaked out on endo-agar. Fractional streaking means that the loop is not streaked over the whole endo-plate but only a single layer of organisms is applied to the surface of medium near the petri dish edge. A second sterile loop is then used to streak part of this material at right angles and over a third of the medium surface. The plate is then turned a further 90° and a newly sterilized loop used to transfer further material to part of the medium not yet streaked. In this way, single organisms can be obtained and can be identified by biochemical tests forming a "colored series".

Colonies suspected of being Escherichia coli are moist and dark red with a gold shimmering metallic tinge. Coliform bacteria grow as moist red colonies with constant or varying metallic effect and with or without slime formation.

A variety of commercially available kits such as API, Enterotube, Titertek etc. may be used for biochemical tests. These consist of pre-prepared culture media which are inoculated with material from a single colony and then incubated.

The procedure and incubation are carried out according to the manufacturer's instructions. The evaluation is often carried out by obtaining a numerical code based on the occurrence of positive or negative metabolic reactions (see below). Reference to a code number in a list provided allows the type of organism to be identified.

Where such kits are not available, the identification media must be prepared, inoculated and the organism identified according to the results obtained.

Determination of biochemical properties:

As a "colored series", the following media are inoculated and incubated for 20 h ± 4 h at the given temperatures (Table 23):

**Table 23:** Determination of biochemical properties on different media of a "colored series"

| Culture medium | Incubation Temperature (°C) | Positive Reaction | Negative Reaction |
|---|---|---|---|
| Nutrient agar plate | 37 | single typical colonies only | morphologically distinguishable colonies |
| Simmons citrate agar angled | 37 | growth with color change from green to blue | no growth no color change |
| Koser citrate solution | 37 | turbidity caused by bacterial growth | no growth, clear solution |

**Table 23:** continued

| Culture medium | Uncubation Temperature (°C) | Positive Reaction | Negative Reaction |
|---|---|---|---|
| Glucose-peptone solution | | turbidity, gas development, color indicator change from purple to yellow | no growth, no gas development, no color change |
| culture A) | 37 | | |
| culture B) | 44 | | |
| Lactose-peptone solution | 44 | turbidity, gas development, color change from purple to yellow | no growth, no gas development, no color change |
| Neutral red mannit boullion | 44 | turbidity, gas development, color change from red to yellow | no growth, no gas development, no color change |
| Urea Kligler Agar | 37 | slant surface: color change from red to yellow, a result of acid production. Prick: gas development, blackening by $H_2S$, $NH_4$-formation by urea degradation | no gas development, no color change through acid production, no blackening by $H_2S$, no $NH_4$ production |
| Tryptophan-Trypton Bouillon | 37 | turbidity growth, red color on addition of indole reagent | no growth, no red color with indole reagent |
| Buffered nutrient solution | 37 | solution is divided into two reagent glasses | |
| | | a) Methylene Red Sample | |
| | | color change from yellow to red | indicator remains yellow |
| | | b) Voges-Proskauer Reaction (addition of KOH + creatinine) | |
| | | red color after 1 - 2 minutes | no color after 2 minutes |
| Nutrient gelatine | 20-22 | liquefaction in inoculation area | no liquefaction |

For a pure culture the nutrient agar plate must show only typical colonies having the same appearance. The presence of different colonies means that the starting colony was of a mixed variety and unsuitable for differentiation purposes. In such cases, sub-cultures must be streaked again on endo-agar. Where a pure culture is present on the nutrient agar plate, the cytochrome oxidase reaction is carried out. 2-3 drops of oxidase reagent N-tetramethyl-p-phenylene-diamine dihydrochloride are placed on the colonies using a dropper. A positive reaction is shown by the colonies changing to violet blue in 1-2 minutes. No color change is visible in the negative case.

A positive cytochrome oxidase reaction shows that E.coli and coliform bacteria are not present, as long as the culture is pure.

For organisms exhibiting a negative cytochrome oxidase test, the "colored series" is to be interpreted according to the following scheme (Table 24):

**Table 24:** Biochemical Properties of E.coli and Coliform Bacteria

|  | E.coli | Enterobacter | Klebsiella | Citrobacter |
|---|---|---|---|---|
| Glucose fermentation at 37°C | + | + | + | + |
| at 44°C | + | +/- | +/- | +/- |
| Lactose fermentation at 44°C | + | +/- | +/- | +/- |
| Mannite fermentation at 44°C | + | +/- | +/- | +/- |
| Citrate degradation | - | + | + | + |
| Indole formation | + | - | - | +/- |
| Methyl red test | + | - | - | + |
| Voges-Proskauer reaction | - | + | + | - |
| Urea degradation | - | +/- | + | +/- |
| $H_2S$ formation | - | - | - | +/- |
| Gelatine liquefaction | - | - | - | - |

+ = positive
- = negative
+/- = differing behavior of various species

Important: E. coli grows at 44 °C and ferments glucose, lactose and mannite with gas production; it forms indole and has a positive methyl red test. Whereas the Voges-Proskauer reaction, urea degradation and hydrogen sulfide production are negative. Citrate is not degraded.

"Coliform" bacteria (i.e. Enterobacter sp.; Klebsiella sp. and Citrobacter sp.) may not grow at 44 °C but at 37 °C and can degrade citrate. They are differentiated among themselves by $H_2S$ formation, urea degradation, indole formation, methyl red reaction and Voges-Proskauer reaction.

Membrane Filter Technique

Larger water volumes (100 mL or more) to be tested are passed through a sterile membrane filter (0.45 μm) which is then incubated in Lactose-peptone solution at 37 °C or placed free of air bubbles on endo-agar or endo nutrient cardboard disks. After incubation, any suspect colonies are further identified as described above. Turbid samples should not be membrane filtered as pore blockage and subsequent inhibition of growth on the filter surface may occur.

The Coli Titre

The so-called coli titer gives the smallest volume of sample in which one bacterium of the Enterobacteriacea is just detectable. It is obtained by incubating a series of different water volumes (e.g. 100 mL, 10 mL, 1 mL, 0.1 mL etc.) in correspondingly concentrated lactose-peptone solutions. The smaller the sample volume in which a reproducible bacterium can be identified, the larger the bacteria concentration in the sample (Table 25).

**Table 25:** Relationship between Coli titer and bacterial density

| Titer (mL) | Bacteria/mL |
|---|---|
| 1000 | 0 |
| 100 | 0 |
| 10 | 0 |
| 1.0 | 1 - 9 |
| 0.1 | 10 - 99 |
| 0.01 | 100 - 999 |
| 0.001 | 1000 - 9999 |

Example of Results:
"Escherichia coli and coliform bacteria are not detectable in 100 mL."
"Escherichia coli is detectable in 0.1 mL water."

6.2.2.3.3 Further Hygienically Important Microorganisms in Water

Faecal Streptococci (Enterococci)

As they are normally present in the intestinal tract of man and animals, these organisms are also indicators of faecal contamination of water. They rarely multiply in water but possess an above average resistance to heat, alkali and salts. Thus, they grow at pH 9.6 and temperatures of 10 -45 °C in a medium with 6.5% NaCl content and are not inhibited by azide.

The microbiological determination can be performed either with the liquid enrichment method in azide-dextrose bouillon or with the aid of membrane filtration. A more accurate final identification is carried out microscopically and by serological methods (Literature list: Suess, WHO-Handbook).

### Pseudomonas aeruginosa

This organism is also a faecal indicator. It is a facultative human pathogen, often plays a role in wound, eye or ear infections and is extremely resistant to antibiotics. It is often found in hospitals and has been isolated from waters passing the E. coli- coliform bacteria test.

The microbiological preparations are performed with the liquid enrichment in malachite green broth followed by transfer inoculation to cetrimide medium or with the aid of membrane filtration, the filter being placed on cetrimide agar. After incubation at 37 C, these organisms form the green pigments fluorescein and pyocyanin and the cultures have a characteristic sweetish aromatic odor.

### Clostridium perfringens

This anaerobic spore forming bacterium is also present in faeces of warm blooded animals but at much lower concentrations than e.g. Escherichia coli.

Since clostridial spores can survive for a long time in the environment, their detection in the absence of Escherichia coli or coliform bacteria points to an older water contamination and is not evidence of a current hygienically important water contamination.

The microbiological detection can be carried out by liquid enrichment in which 20 - 100 mL samples are mixed with an equal amount of doubly concentrated dextrose-iron citrate-sodium sulfite broth incubated under anaerobic conditions. Blackening of the liquid medium indicates a positive reaction.

In the membrane filter method, the water is drawn through the filter. The reverse side is placed on dextrose-iron citrate-sodium sulfate agar and incubation carried out anaerobically at +37 °C. Black colonies indicate a positive reaction.

### Parasites, Especially Worm Eggs

Waste water and, in particular, unprocessed waste water present a source of danger for the transfer of worm parasites to man and animals. Worm eggs, larvae and other developmental stages can reach water via faeces from humans and domestic animals. The efficiency of eradication of worm eggs in a waste water purification plant must therefore be determined before and after processing. Worm eggs can survive in sewage sludge even through the rotting stage so that its use as fertilizer may present a risk of infection. A particular risk is in the application of unprocessed water or dung to fruit and vegetables.

Transfer of worm diseases is also possible in swimming baths.

In water treatment plants fed by surface waters, a regular examination for parasite eggs is recommended, especially when cattle are reared near to these surface waters. Parasitic worm eggs are common throughout the world and can cause several potentially fatal diseases.

### Enrichment of Worm Eggs in Saturated Sodium Chloride

A sludge or water sample weighing approx. 1 g is mixed 1 : 27 with sodium chloride solution. The latter is completely saturated (approx. 377 g sodium chloride to 1 L water) and is added slowly, initially dropwise with continuous stirring so that a fine even bilayer results which is then left still for 15 - 20 minutes. The eggs of all nematodes and certain cestodes (Taenia, Hymenolepis) float in saturated sodium chloride owing to their lower specific gravity in contact with other sludge components which sink. Small beakers are employed to concentrate the rising worm eggs over a small area. A wire loop (1 cm diameter) bent at right angles to the stem and placed flat on the surface is used to remove worm eggs without contaminating particles. The drop is placed on a slide and examined microscopically without a cover glass. This method is unsuitable for the detection of trematode eggs, the eggs of Diphylobothrium and the unfertilized eggs of Ascaris. It must be taken into account that Trichuris eggs rise comparatively slowly. With negative results but continuing suspicion, it is recommended that the tests be repeated.

The recovery can be significantly increased, even for trematode eggs, by employing a saturated solution of zinc chloride owing to its higher specific gravity.

#### 6.2.2.4 Preparation of Nutrient Solutions and Media

General:

For the microbiological examination of water and determination of total bacterial count, gelatine and agar agar culture media are employed. In addition, liquid nutrient solutions are used as enrichment media and in the "colored series" of biochemical tests for differentiation of the Enterobacteriaceae.

Gelatine and agar agar are employed as binders for the nutrients and by providing a solid medium allow the organisms to grow separately and in fixed positions i.e. assist colony formation. Gelatine is a high molecular weight protein which is commercially available. Gelatine culture media liquefy at temperatures above 25 °C so that incubation must be carried out at lower temperatures e.g. at 20-22 °C. Agar agar is a polysaccharide sulfuric ester and is obtained from certain red algae. It is also available commercially. Agar culture media liquefy at approx. 100 °C and solidify again at below 45 °C. Culture media incorporating agar may be incubated at temperatures above 45 °C after solidifying once.

In the case of gelatine culture media, it must be considered that certain bacteria and moulds produce protein cleaving enzymes which can attack the gelatine and cause liquefaction of the media. Proteolytic bacteria (mainly Pseudomonas species) occur mainly in surface waters so that an increased presence of gelatine-liquefying organism in deep water indicates a possible influence of surface water.

Maintenance of optimum pH in the culture medium is of great importance for the later growth. It is therefore necessary to make provision for pH regulation during the preparation.

1 M HCl is employed to reduce excessively high values and a 10 % soda solution or 1 M NaOH to increase them.

The following instructions are suitable for preparing the culture media employed in the microbiological testing of water:

<u>Gelatine Culture Media:</u>

| | |
|---|---|
| Meat extract | 10 g |
| Peptone | 10 g |
| Sodium chloride | 5 g |
| Gelatine | 120 - 150 g |

(The gelatine content should be somewhat higher in warmer seasons than in cold ones).

| | |
|---|---|
| Deionized water | 1000 mL |

Preparation:

1000 mL of deionized water are poured on to the given amounts of meat extract, peptone, sodium chloride and gelatine contained in a 2000 mL Erlenmeyer flask. The gelatine is allowed to swell for 1 hour at about 25 °C and is then dissolved on a water bath at about 50 °C. Stronger heat is to be avoided before pH adjustment as gelatine normally reacts as an acid so that at higher temperatures the protein can coagulate. After dissolution of the gelatine, the medium is adjusted to pH 7.2 and the beaten whites of two hens eggs added for clarification.

After thorough mixing, the liquid is heated with steam for 30-45 minutes at 100 °C during which the coagulated egg protein precipitates. The clear nutrient medium is poured on to a wettened folded filter paper and filtered through a hot water funnel or steam pot. The first part of the filtrate is filtered again until the medium runs clearly. A gelatine culture medium should be completely clear and have a yellowish color. 10 mL portions of the filtered medium are filled into test tubes, closed with a cotton wool wad, cellulose stopper or metal cap and fractionally sterilized in a steam sterilizer (3 x 20 minutes at 24 hour intervals).

Excessive heating of gelatine media must be avoided as the gelatine can lose its ability to solidify.

<u>Nutrient Agar:</u>

| | |
|---|---|
| Meat extract | 10 g |
| Peptone | 10 g |
| Sodium chloride | 5 g |
| Agar agar | 30 g |
| Deionized water | 1000 mL |

Preparation:

1000 mL of deionized water are poured on to the given amounts of meat extract, peptone, sodium chloride and agar agar in a 2000 mL Erlenmeyer flask. The mixture is allowed to swell at 25 °C and then dissolved by heating in steam. The pH value is adjusted to 7.2 - 7.5 by careful addition of soda solution or sodium hydroxide solution to the hot liquid medium. If necessary, clarification with protein is carried out as for gelatine medium. The liquid medium is then filled into 10 mL test tubes which are closed with cotton wool wads, cellulose stoppers or metal caps. Sterilization is carried out either in the autoclave by heating at 121 °C for 15 minutes or fractionally in a steam pot (3 x 30 minutes at 24 hour intervals).

The agar culture medium melts at 100 °C and solidifies at 45 °C. It can therefore be employed for cultivation of heat tolerant and thermophilic microorganisms.

Lactose-Peptone- Solution:

| | |
|---|---|
| Lactose | 20 g |
| Peptone | 20 g |
| Sodium chloride | 10 g |
| Deionized water | 1000 mL |
| Bromocresol purple indicator | (1 g bromocresol purple is dissolved in 100 mL demionized water) |

Preparation:

The given amounts of peptone and sodium chloride are dissolved in 1000 mL deionized water by heating on a steam bath. After leaving for about 1 hour on the steam bath, the given amount of lactose is added and the mixture heated for a further 20 minutes. The pH is adjusted to 7 by addition of soda or sodium hydroxide solution and 2 mL of the bromocresol purple indicator is added. This doubly concentrated solution is poured into culture dishes in 10 mL or 100 mL amounts (for determination of the coli titre according to the enrichment method) and sterilized for 20 minutes in the autoclave at 121 °C.

The normal concentration of lactose-peptose solution is prepared by thinning the nutrient solution as prepared above with the same volume of deionized water prior to the addition of the bromocresol purple indicator. The pH is then adjusted and indicator solution added. 10 mL portions are filled into test tubes together with a Durham tube which are then sterilized for 30 minutes in an autoclave at 121 °C.

Durham tubes are similar to 4-5 cm long test tubes with 6-8 mm diameter and are placed openend downwards in the test tubes. During sterilization the air escapes from the Durham tube which fills with liquid. Any gas formed during incubation of the inoculated solution will collect in the Durham tube.

Endo-Agar (Lactose-fuchsin-sulfite agar)

| | |
|---|---|
| Nutrient agar | 1000 mL |
| Lactose | 15 g |

|  |  |
|---|---|
| Concentrated alcoholic fuchsin solution | 5 mL (10 g diamond fuchsin dissolved in 90 mL ethanol) |
| Sodium sulfite solution, 10 % | approx. 25 mL (10 g sodium sulfite, $Na_2SO_3 \cdot 7\,H_2O$, dissolved in 90 mL deionized water) |

Preparation:

The given amounts of lactose and fuchsin solution are added to 1000 mL of nutrient agar which has been liquified by heating on a steam bath and the liquid is well mixed. The culture medium thus attains an intense red color and is then decolorised by addition of sodium sulfite solution. This addition must be carried out very carefully until the hot medium becomes just weakly pink (normally about 25 mL of sodium sulfite solution are required). When cold, the medium is then almost colorless. This should be checked by taking a sample of the hot medium in a test tube and cooling in a stream of water until solid.

The endo-agar thus prepared contains 3 % agar agar and is suitable for making sub-cultures by streaking out. The medium is sensitive to light and must be stored cold and in the dark.

For preparation of endo-agar cultures for use with membrane filters, a smaller concentration of agar agar is desired in order to allow better diffusion of nutrient and indicator. A nutrient agar containing only 1 % agar agar (10 g in 1 L) is therefore used and must be prepared before making the special endo-agar. The endo-agar with 1 % agar-agar is too soft for making subcultures by streaking out as the surface is easily damaged by the platinium loop or needle.

The endo-agar culture medium is not put into tubes but in Erlenmeyer flasks which are fractionally sterilized for 3 x 20 minutes at 24 hour intervals before pouring the medium into petri dishes. The pouring must be carried out in a darkened room allowing the medium to solidify in the dark. In the absence of such a room, the plates must be covered with opaque material during solidification. Before use, the endo-agar is stored cold and in the dark and is pre-dried at 37 °C in the incubator before inoculation. For drying, the lid and base of the petri dish are placed face down in the incubator for about 30 minutes at 37 °C.

Culture Media for Selective Detection of Enterococci (according to Slanetz und Bartley):

|  |  |
|---|---|
| Tryptose | 20 g |
| Yeast extract | 5 g |
| Glucose | 4 g |
| Disodium Hydrogen Phosphate ($Na_2HPO_4 \cdot 2\,H_2O$) | 4 g |
| Sodium azide ($NaN_3$) | 0.4 g |
| Triphenyltetrazolium chloride (TTC) | 0.1 g |
| Agar-agar | 10 g |

|  |  |
|---|---|
| Deionized water | 1000 mL |

Preparation:

The above ingredients with exception of sodium azide and TTC are dissolved in the deionized water by careful heating to boiling. After cooling to 46 °C, the solution is treated with 10 mL of a 1 % aqueous sterile filtered TTC solution and 4 mL of a 10 % aqueous sterile filtered sodium azide solution. No further sterilisation is carried out. The medium is immediately poured into sterile petri dishes and allowed to solidify on a horizontal surface. These culture media are especially suitable for water tests by the membrane filter method but are only stable for a limited time.

### Culture Medium for Selective Detection of Pseudomonas aeruginosa
Cetrimide Agar:

| | |
|---|---|
| Peptone | 20 g |
| Magnesium chloride ($MgCl_2$) | 1.4 g |
| Potassium sulfate ($K_2SO_4$) | 10 g |
| N-acetyl-N,N,N-trimethyl-ammonium bromide | 0.5 g |
| Agar-agar | 13.6 g |
| Glycerine, doubly distilled | 10 mL |
| Deionized water | 1000 mL |

Preparation:

1000 mL deionized water is rigorously shaken together with the listed ingredients (with exception of glycerine) in a glass flask. After a swelling period of 10 - 30 minutes at 20-25 °C during which the nutrient mixture must be kept still, 10 mL of doubly distilled glycerine are added and the mixture heated to boiling with frequent swirling. The pH value is then adjusted between 7.3 and 7.4 and the selective agar is either dispensed in 10 mL volumes in test tubes which are sterilized for 15 minutes in the autoclave at 121 °C, or the whole amount is sterilized in the flask and poured into sterile petri dishes before solidification.

Ready Made Culture Media:

Dried culture media in powder form containing all components necessary for cultivating the respective microorganisms are readily available commercially. According to the delivered instructions, an amount of the powder must be weighed out and treated with a certain volume of deionized water or mains water. The medium is then generally dissolved by heating and the pH adjusted. After putting into flasks or test tubes, the medium is sterilized either in an autoclave or fractionally in a steam bath. Adjustment of the pH is always recommended. Some manufacturers also deliver the ready-made culture media in tablet form. The tablets correspond to a particular amount of water so that weighing is not necessary.

### 6.2.3 Specialized Soil Analyses

Although the present handbook is concerned mainly with water and waste water, a description of some important methods of soil analysis is also considered to be relevant. Investigations of fertility and melioration of soil are usually closely related to questions of water characteristics and economy. However, detailed explanations based on soil science will not be given, the emphasis being placed here on practical aspects of the analysis.

### 6.2.3.1 Mesh Size

The grain size analysis should reveal the proportions of various particle sizes as this strongly influences the physical properties of the soil.

#### Apparatus
Sieve set (pore sizes 2 mm; 0.63 mm; 0.2 mm; 0.063 mm)
Measuring cylinder, 1000 mL
Weighing glasses

#### Reagents and Solutions
100 g sodium diphosphate ($Na_4P_2O_7$) is dissolved in water and made up to 1 L.

#### Sample Preparation
The dried and roughly crushed raw soil is passed through a sieve (2 mm), the proportion of stones and roots being separately weighed. The portion of fine soil (< 2 mm) is further treated.

#### Measurement
20 g air-dried fine soil are left for approx. 8 hours with 25 mL sodium diphosphate solution. 200 mL water are then added and the mixture shaken. The suspension is then passed through a sieve set directly into the 1 L measuring cylinder. The individual sieves are washed until the finer particles have passed the relevant sieve. The set is then dried at 105 °C in a drying oven and the individual sieves weighed.

The suspension in the measuring cylinder is made up to 1000 mL with water and thoroughly shaken. After approx. 9 1/2 minutes, 10 mL are pipetted from 20 cm below the surface and transferred to a weighing glass. This suspension contains the fraction smaller than 20 $\mu$m. The fraction smaller than 10 $\mu$m is obtained by pipetting 10 mL suspension after 18 1/2 minutes from the area 10 cm beneath the surface and transferring to a weighing glass. Finally, the clay fraction < 2 $\mu$m is obtained after 3 hours and five minutes from a depth of 4 cm.

All samples thus obtained are dried at 105 °C in the laboratory oven and then weighed. The 25 mg proportion of sodium diphosphate is subtracted from the amounts obtained.

#### Interfering Factors
The method depends on the sedimentation rate of spherical particles in water. The more the particle shape deviates from the ideal spherical form, the smaller is the sedimentation rate so

that such particles will be assigned to a fraction of larger particle size than their mean diameter.

Calculation of Results
The weights of the individual fractions can be related to the raw soil or the fine soil depending on requirements. The weights of the dried fractions, corrected for sodium diphosphate content, are multiplied by 5 giving the portion of each fraction as % of the original weight.

### 6.2.3.2 Hydrolytic Acidity (H-Value)

Determination of the total acid content of soil involves determination of the free $H^+$ ions as well as the complex bound $H^+$ ions. Indirect methods are mainly employed in which hydrolyzable salts are brought into contact with the soil and the free acid formed by reaction of sorbed $H^+$ ions with the anion is measured by titration.

The value obtained is multiplied by empirical factors to estimate the total acid.

Apparatus
Titration equipment

Reagents and Solutions
Calcium acetate solution:  88.09 g calcium acetate $(CH_3COO)_2Ca \cdot H_2O$ are dissolved and made up to 1 L with water. The color is adjusted to pink with 0.1 M NaOH and a few drops of phenolphthalein.

Measurement
100 g air-dried fine soil are treated with 250 mL calcium acetate solution and shaken for 1 hour. After filtration, the first 30 mL are discarded and 125 mL of filtrate are titrated against phenolphthalein to give a pink color.

Calculation of Results
The consumed sodium hydroxide is multiplied by empirical factors to yield values for the total soil acidity:

| | |
|---|---|
| For adjustment to pH 7 | F = 1.5 |
| For adjustment to pH 7.5 | F = 2.0 |
| For adjustment to pH 8.0 | F = 2.5 |
| For adjustment to pH 8.5 | F = 3.25 |

If titration is performed against phenolphthalein, the factor F = 3.25 is used.

The hydrolytic acidity is given in mL of 0.1 M sodium hydroxide per 100 mL soil. The total acidity in mL of 0.1 M sodium hydroxide is then calculated as follows:

## 6.2 Analytical Methods

Total acidity (in mL of 0.1 M NaOH) = $2 \cdot x \cdot 3.25$

x = consumed 0.1 M NaOH (ml)

The sum of total acid and base saturation (S-value; section 6.2.3.3) gives the total base saturation (= exchange capacity) in mmol/100 g soil.

### 6.2.3.3 Exchangeable Basic Substances (S-value)

In determining the s-value, the fixed cations such as calcium, magnesium, sodium, potassium and ammonium which are fixed to negatively charged soil colloids are covered but the carbonates are not.

The s-value is important for estimation of the ability to deliver bases and therefore for the soil fertility.

#### Apparatus
Titration equipment

#### Reagents and Solutions
Hydrochloric acid 0.1 M
Sodium hydroxide 0.1 M
Sodium potassium tartrate solution ($C_4H_4O_6KNa \cdot 4 H_2O$), 10 %
Ethyl alcohol - water mixture (60 : 40 v/v)

#### Sample Preparation
15 to 25 g of moist fine earth are placed on a folded filter paper and several times the alcohol-water mixture is poured over it in order to displace the soil water but without occurrence of exchange processes. The residue is dried with the filter paper at 105 °C in a laboratory oven.

#### Measurement
Depending on the chalk or limestone content of the soil, between 1 and 10 g of air-dried fine earth are weighed out (1 g at 30 to 50 % $CaCO_3$, 10 g at up to 5 % $CaCO_3$), treated with 100 mL 0.1 M HCl and shaken for one hour. After filtration, 20 mL of filtrate are treated with 5 mL of sodium potassium tartrate solution (10 %) and titrated with 0.1 M NaOH against phenolphthalein to give a pink color. The blank consists of 20 mL of 0.1 M HCl and 5 mL sodium potassium tartrate solution, titrated in the same way.

#### Calculation of Results
The difference between the titration values for the blank and the soil extract is multiplied by a factor 50 for an initial weight of 1 g (factor 5 for weight of 10 g), thereby giving an S-value in mol/100 g soil. This value includes carbonate so that a carbonate determination must be carried out in parallel. The result of the s-value titration is then corrected for the carbonate

content of the soil.

### 6.2.3.4 Exchange Capacity

The exchange capacity of soil is a measure of the proportion of colloidal substances whose surfaces can function as cation exchangers. During the determination, the fixed cations are replaced by the action of a high concentration of easily exchangeable cations. The amount of exchanged cations corresponds to the exchange capacity of the soil. The values for mineral soils lie between 15 and 40 mol/100 g soil and for strongly humus soils up to 300 mol/100 g soil.

#### Apparatus
Distillation equipment as described in 6.2.1.11

#### Reagent and Solutions
Ammonium oxalate solution 0.2 M
Calcium carbonate
Activated charcoal, powdered

#### Measurement
250 mL of 0.2 M ammonium oxalate solution are added to 20 g of air-dried fine earth together with 5 g of activated charcoal (for fixation of ammonium humates) and 0.5 g of calcium carbonate (as buffer). The mixture is then shaken for 2 hours and filtered. 2 drops of conc. sulfuric acid are added to the filtrate. The procedure for determining the ammonium content after distillation as described in 6.2.1.11 is then performed on a 25 mL aliquot of this solution. A blank sample consisting of 25 mL of 0.2 M ammonium oxalate solution is treated in the same way.

#### Calculation of Results
The difference between the ammonium content of the blank and the soil extract gives the amount of $NH_4^+$ ions taken up by the soil. The exchange capacity is given in mol/100 g soil.

### 6.2.3.5 Carbonate Content

Knowledge of the calcium carbonate content of a soil is of considerable importance as this parameter influences e.g. the soil texture and permeability. Also, chemical processes in the soil are affected by the carbonate content.

The volumetric method of determination will be described here.

#### Apparatus
Measuring apparatus according to the Figure 44.

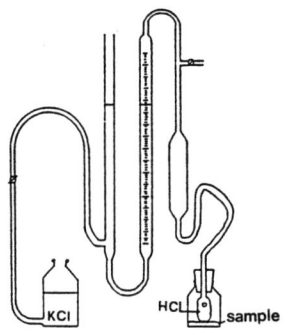

**Fig. 44:** Instrument for the determination of carbonate in soil

Reagents and Solutions
Hydrochloric acid, 10 %
Potassium chloride solution, 2 %

Measurement
Depending on results of a qualitative test sample with hydrochloric acid, between 2 and 10 g of air-dried fine soil are placed in the gas generator and the insert filled with 20 ml hydrochloric acid (10%). After connecting to the apparatus, the graduated tube is filled by raising the level container. The gas generator is then tilted so that the hydrochloric acid makes contact with the floor. Depending on the level set in the graduated tube, a pressure compensation is attained by raising or sinking the level container so that, after approx. 10 minutes, the gas volume can be read off.

Interfering Factors
The measurement is relatively inaccurate for a carbonate content under 1 %. The presence of magnesium and iron carbonates leads to lower values owing to the low speed at which both compounds react as well as the changed stoichiometric factors.

Calculation of Results
The carbonate content of soil (as $CaCO_3$) is calculated as follows:

$$CaCO_3 \text{ (\%)} = \frac{V \cdot P \cdot 0.12}{(273 + t) \cdot W}$$

V = measured volume of $CO_2$ (mL)
P = air pressure (hPa = mbar)
t = room temperature (°C)
W = weight of soil (g)

### 6.2.3.6 SAR-Value (Sodium Absorption Ratio)

A high sodium content of soil in relation to the content of other cations (especially calcium

## 6 Laboratory Measurements

and magnesium) means a danger of oversalting and therefore lower profitability.

A simple characterization of the problems caused by sodium may be made by the SAR value. In addition to soil, the determination in irrigation water can be of importance.

Measurement

A soil paste is prepared by adding sufficient water to approx. 250 g moist or air-dried soil with stirring so that it is saturated without the presence of free water. After leaving for 2 hours, the paste is transferred to a suction filter and the water drawn off. The parameters sodium, calcium and magnesium are measured in the soil solution.

Calculation of Results

The SAR-value is calculated as follows:

$$SAR = Na^+ / \sqrt{(Ca^{2+} + Mg^{2+})/2}$$

The concentration is given in millequivalents. A nomogram (Figure 45) simplifies determination of the SAR value. For this purpose, the measured $Na^+$ concentration on ordinate I is connected with the value for the sum of calcium and magnesium concentrations on ordinate II and the SAR value read off from the diagonal. The nomogram for sprinkler water is shown in Figure 46.

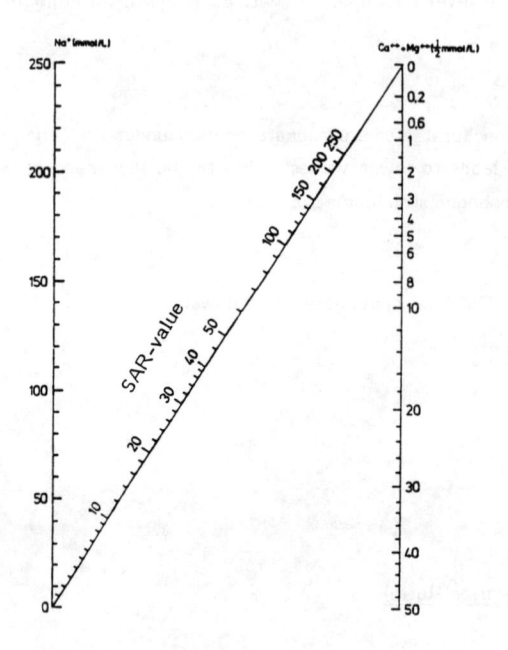

**Fig. 45:** Nomogram for the determination of the SAR-value in soil extracts

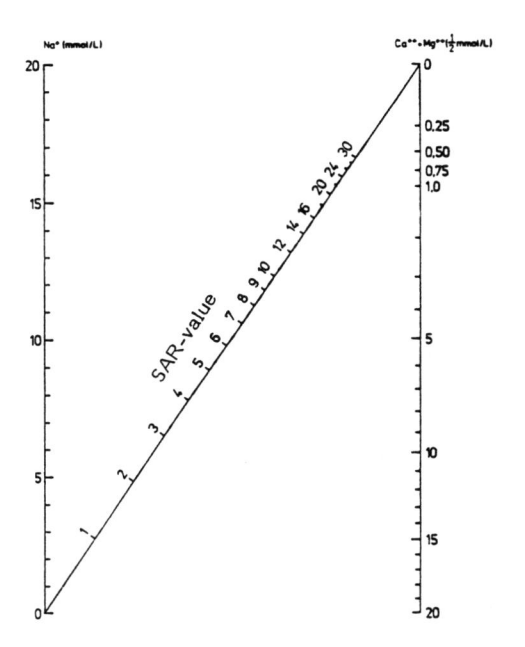

**Fig. 46:** Nomogram for the determination of the SAR-value in sprinkler water

# 7 Interpretation of Test Results

The evalution and interpretation of water and soil test results requires considerable practical experience. In addition, national and international guidelines and tables can be of great assistance. Even taking into account the weaknesses inherent in such tables introduced by the problems of threshold judgement and analytical errors, considerable experience from the areas of toxicology, nutrition, agriculture and technology is contained within them. However, threshold values can never be simply accepted. The local conditions, quality of analytical methods and the test requirements must be included in the interpretation.

The guidelines, recommendations and threshold values described below comprise only a selection and it cannot be claimed that the list is exhaustive.

## 7.1 Ground Water

It has already been discussed in section 3.1 that the evaluation of chemical and bacteriological test results on ground water only makes sense if the natural background values are taken into account. The distinction between natural and man-made contamination is often difficult so that comparative measurements have to be relied upon. Also, temporal variations in concentration must be considered. Since ground water may be utilized in the same way as surface water as a source of drinking water or for irrigation, surface water guidelines should be employed in the evaluation.

## 7.2 Surface Water

Depending on the intended use, various official guidelines for surface water have been drawn up e.g. requirements for bathing waters (EEC Guidelines) or requirements for fishery waters (Federal Republic of Germany). In addition, Germany lists the minimum requirements for waterways in relation to the so-called water classes II/III. Depending on the local conditions found in a particular state (climate, geology), the listed values are more or less suitable for the interpretation of test data.

**Table 26:** Minimum requirements for flowing water systems in the FRG (Basis: Water Classes II/III)

| Parameter | | Minimum requirement | Purification necessary |
|---|---|---|---|
| max. temperature (°C) | | | |
| a) cool summer waters | | 25 | 25 |
| b) warm summer waters | | 28 | 28 |
| oxygen | mg/L | 4 | 4 |
| pH | | 6 - 9 | 6 - 9 |
| ammonium ($NH_4^+$-N) | mg/L | 1 | 2 |
| BOD without nitrification inhibition | mg/L | 7 | 10 |
| COD | mg/L | 20 | 30 |
| phosphorus | mg/L | 0.4 | 1 |
| iron | mg/L | 2 | 3 |
| zinc | mg/L | 1 | 1.5 |
| copper | mg/L | 0.05 | 0.06 |
| chromium | mg/L | 0.07 | 0.1 |
| nickel | mg/L | 0.05 | 0.07 |

**Table 27:** Quality requirements for fishery waters in the FRG (Basis: German Water Class II)

| Parameter | | Salmonide waters | Cyprinide waters |
|---|---|---|---|
| max temperature (°C) | | | |
| a) cool summer waters | | 20 | 25 |
| b) warm summer waters | | 20 | 28 |
| oxygen | mg/L | 6 | 4 |
| pH | | 6.5 - 8.5 | 6.5 - 8.5 |
| ammonium ($NH_4$-N) | mg/L | 1 | 1 |
| BOD without inhibition of nitrification | mg/L | 6 | 6 |
| COD | mg/L | 20 | 20 |
| iron | mg/L | 2 | 2 |
| zinc | mg/L | | |
| a) with 4 mg/L Ca | | 0.03 | 0.3 |
| b) with 20 mg/L Ca | | 0.2 | 0.7 |
| c) with 40 mg/L Ca | | 0.3 | 1 |
| dissolved copper | mg/L | | |
| a) with 4 mg/L Ca | | 0.005 | 0.005 |
| b) with 20 mg/L Ca | | 0.022 | 0.022 |
| c) with 40 mg/L Ca | | 0.04 | 0.04 |
| nitrite ($NO_2^-$ - N) | mg/L | 0.015 | 0.015 |

**Table 28:** Quality requirements for bathing waters (EEC Guidelines; 1975)

| Parameter | | normal value | limit |
|---|---|---|---|
| **Microbiological parameters** | | | |
| total coliform/100 mL | | 500 | 10 000 |
| faecal coliform/100 mL | | 100 | 2 000 |
| faecal streptococcus/100 mL | | 100 | - |
| salmonellae/100 mL | | - | 0 |
| intestinal virus PFU/10 L | | - | 0 |
| **Physico-chemical parameters** | | | |
| pH | | - | 6 - 9 |
| transparency, m | | 2 | 1 |
| color | | - | no abnormal change |
| mineral oils | mg/L | - (< 0.3) | no visible film |
| anionic surfactants | mg/L | - (< 0.3) | no foam formation |
| phenol index | mg/L | 0.005 | 0.005 |
| tar residues, suspended particles | | none | - |

## 7.3 Drinking Water

The quality of drinking water is regulated in most countries by recommendations or legal requirements. Of special importance for the developing countries are the World Health Organisation (WHO) Recommendations of 1984 (Table 29). Furthermore, the EEC Recommendations of 1980 have been assimilated, at least in part, into the national regulations of various Western European countries and should therefore be given attention (Table 30). The German Drinking Water Regulations of 1986 will be presented here as an example (Table 31).

Where surface water is to be used for preparing drinking water, the threshold values from the relevant EEC Guidelines (1975) may be of importance (Table 32). For questions relating to the addition of substances during drinking water processing, the draft of German drinking water processing regulations (as of 1987) may be consulted (Table 33).

**Table 29:** WHO recommendations for drinking waters quality (extract)

### I Microbiological parameters

| Organism | Normal value (count/100 mL) | Remarks |
|---|---|---|
| **A.** Water distribution via mains | | |
| A.1 Treated water fed into mains | | |
| E. coli | 0 | turbidity <1 NTU; pH-value on chlorination <8.0; after 30 min contact time free chlorine 0.2 - 0.5 mg/L |
| Coliforms | 0 | |
| A.2 Untreated water fed into mains | | |
| E. coli | 0 | in 98% of samples tested per year in larger supply systems |
| Coliforms | 0 | |
| Coliforms | 3 | occasionally but not in sequential samples |
| A.3 Mains water | | |
| E. coli | 0 | in 95% of samples tested per year in larger supply systems |
| Coliforms | 0 | |
| Coliforms | 3 | occasionally but not in sequential samples |
| **B** Non-mains water supply | | |
| E. coli | 0 | |
| Coliforms | 10 | should not appear in sequential samples |

### II Chemical and sensory parameters

| Parameter | Normal value | Remarks |
|---|---|---|
| a) | | |
| color (TCU) | 15 | |
| taste and odor | none | |
| turbidity (NTU) | 5 | if possible 1 on disinfection |
| pH | 6.8 - 8.5 | |
| dissolved solids (mg/L) | 1 000 | |
| hardness as $CaCO_3$ (mg/L) | 500 | |
| chloride (mg/L) | 250 | |
| sulfate (mg/L) | 400 | |
| hydrogen sulfide | not detected | |
| sodium (mg/L) | 200 | |
| iron (mg/L) | 0.3 | |
| manganese (mg/L) | 0.1 | |
| aluminum (mg/L) | 0.2 | |

**Table 29:** continued

| Parameter | Normal value | Remarks |
|---|---|---|
| copper (mg/L) | 1 | |
| zinc (mg/L) | 5 | |
| b) | | |
| arsenic (mg/L) | 0.05 | |
| cadmium (mg/L) | 0.005 | |
| chromium (mg/L) | 0.05 | |
| cyanide (mg/L) | 0.1 | |
| fluoride (mg/L) | 1.5 | |
| lead (mg/L) | 0.05 | |
| mercury (mg/L) | 0.001 | |
| nitrate (mg/L) | 45 | |
| selenium (mg/L) | 0.01 | |
| c) | | |
| aldrin, dieldrin ($\mu$g/L) | 0.03 | |
| benzene ($\mu$g/L) | 10 | |
| benzo(a)pyrene ($\mu$g/L) | 0.01 | |
| carbon tetrachloride ($\mu$g/L) | 3 | |
| chlordane ($\mu$g/L) | 0.3 | |
| chloroform ($\mu$g/L) | 30 | |
| 2,4 D ($\mu$g/L) | 100 | |
| DDT ($\mu$g/L) | 1 | |
| 1,2 dichloroethene ($\mu$g/L) | 10 | |
| 1,1 dichloroethane ($\mu$g/L) | 0.3 | |
| heptachlor and heptachloroepoxide ($\mu$g/L) | 0.1 | |
| $\gamma$-HCH ($\mu$g/L) | 3 | |
| methoxychlor ($\mu$g/L) | 30 | |
| pentachlorophenol ($\mu$g/L) | 10 | |
| tetrachloroethene ($\mu$g/L) | 10 | |
| trichloroethene ($\mu$g/L) | 30 | |
| 2,4,6-trichlorophenol ($\mu$g/L) | 10 | |

In 1980, an EEC Guideline concerned with the quality of water for human consumption was drawn up. It contains a multitude of recommended parameter values and maximum permissible concentrations, but for certain parameters values remain to be agreed (Table 30).

## 7.3 Drinking Water

**Table 30:** EEC Guidelines concerning the quality of water for human consumption

| Parameter | Normal value | max. permissible concentration | Remarks |
|---|---|---|---|
| **A. Organoleptic parameters** | | | |
| color (mg/L Pt/Co) | 1 | 20 | |
| turbidity (mg/L $SiO_2$) | 1 | 10 | or measurement of transparency by the Secchi-disc |
| threshold level for odor (dilution factor) | | 2 at 12 °C  3 at 25 °C | |
| threshold level for taste (dilution factor) | | 2 at 12 °C  3 at 25 °C | |
| **B. Physico-chemical parameters** (in connection with the natural composition) | | | |
| temperature (°C) | 12 | 25 | |
| pH | 6.5 - 8.5 | | |
| electrical conductivity ($\mu S\ cm^{-1}$) (20 °C) | 400 | | |
| chloride (mg/L) | 25 | | effects above 200 mg/L |
| sulfate (mg/L) | 25 | 250 | |
| calcium (mg/L) | 100 | | |
| magnesium (mg/L) | 30 | 50 | |
| sodium (mg/L) | 20 | 175 | will be decreased in the future |
| potassium (mg/L) | 10 | 12 | |
| aluminum (mg/L) | 0.05 | 0.2 | |
| evaporation residue (mg/L) | | 1500 | |
| **C. Parameters for undesirable substances** | | | |
| nitrate (mg/L) | 25 | 50 | |
| nitrite (mg/L) | | 0.1 | |
| ammonium (mg/L) | 0.05 | 0.5 | |
| Kjeldahl-N (mg/L) | | 1 | |
| oxidizability $KMnO_4$ (mg/L) | 2 | 5 | |
| hydrogen sulfide | | not detectable sensorily | |
| $CHCl_3$ extractable substances (mg/L) | 0.1 | | |
| petrol ether extractable hydrocarbons (mg/L) | | 0.01 | |
| phenol index (mg/L) | | 0.0005 | excluding the natural phenols which do not react with Cl |

**Table 30:** continued

| Parameter | Normal value | max. permissible concentration | Remarks |
|---|---|---|---|
| boron (mg/L) | 1 | | |
| anionic surfactants (laurylsulfate)(mg/L) | | 0.2 | |
| organic chlorine compounds (not pesticides)(mg/L) | 0.001 | 0.025 | |
| iron (mg/L) | 0.05 | 0.2 | |
| manganese (mg/L) | 0.02 | 0.05 | |
| copper (mg/L) | 0.1 | | after 12 h stay in mains network: 3 mg/L |
| zinc (mg/L) | | 0.1 | after 12 h stay in mains network: 5 mg/L |
| phosphorus (as $P_2O_5$) (mg/L) | 0.4 | 5 | |
| fluoride (mg/L) | | 1.5 (8–12 °C) 0.7 (25–30 °C) | |
| barium (mg/L) | 0.1 | | |
| silver (mg/L) | | 0.01 | in certain cases 0.08 |

**D. Parameter for toxic substances**

| Parameter | | max. permissible concentration | Remarks |
|---|---|---|---|
| arsenic (mg/L) | | 0.05 | |
| cadmium (mg/L) | | 0.005 | |
| cyanides (mg/L) | | 0.05 | |
| chromium (mg/L) | | 0.05 | |
| mercury (mg/L) | | 0.001 | |
| nickel (mg/L) | | 0.05 | |
| lead (mg/L) | | 0.05 | |
| antimony (mg/L) | | 0.01 | |
| selenium (mg/L) | | 0.01 | |
| pesticides (mg/L) | | 0.0001 each substance 0.0005 total | |
| polycyclic aromatic hydrocarbons (PAH) (mg/L) | | 0.0002 | 6 reference compounds |

**E. Microbiological parameters**

| Parameter | | max. permissible concentration | Remarks |
|---|---|---|---|
| colony count at 37 °C (count/mL) | | 10 | |
| colony count at 22 °C (count/mL) | | 100 | |
| E. coli (count/100 mL) | | 0 1 | membrane filter method titer determination |
| coliforms | | 0 1 | membrane filter method titer determination |

**Table 30:** continued

| Parameter | Normal value | max. permissible concentration | Remarks |
|---|---|---|---|
| faecal streptococci |  | 0 | membrane filter method |
|  |  | 1 | titer determination |
| sulfite-reducing Clostridium (count/20 mL) |  | 1 | titer determination |

The EEC Guidelines on quality of water for human consumption (1980) initiated their inclusion in the national laws of the member countries. This process has the aim of leading to uniform requirements within the EEC for the properties, testing and assessment of drinking water.

In Table 31, the threshold values from the German drinking water regulations (1986) are shown. They are also applicable to water used in the food industry.

**Table 31:** Extract of drinking water regulations (Federal Republic of Germany, 1986)

| Parameter | limit |
|---|---|
| **I. Limits for chemical substances** | |
| aluminum (mg/L) | 0.2 |
| ammonium (mg/L) | 0.5 |
| arsenic (mg/L) | 0.04 |
| lead (mg/L) | 0.04 |
| cadmium (mg/L) | 0.005 |
| total chromium (mg/L) | 0.05 |
| cyanide (mg/L) | 0.05 |
| fluoride (mg/L) | 1.5 |
| iron (mg/L) | 0.2 |
| magnesium (mg/L) | 50 |
| nickel (mg/L) | 0.05 |
| nitrate (mg/L) | 50 |
| nitrite (mg/L) | 0.1 |
| mercury (mg/L) | 0.001 |
| sodium (mg/L) | 150 |
| sulfate (mg/L) | 240 (except in sulfate-containing soil) |
| surfactants (mg/L) | 0.2 |
| polycyclic aromatic hydrocarbons (as C) (mg/L) | 0.0002 |
| volatile halogenated compounds (mg/L) (sum of 1,1,1-trichloroethane trichloroethene, tetrachloroethene, dichloromethane) | 0.025 |
| carbon tetrachloride (mg/L) | 0.003 |

**Table 31:** continued

### II. Microbiological Parameters

The threshold values are the same as those listed in Section E of Table 30. In addition, disinfected drinking water should not exceed the counts of 20/mL at 20 °C and 2/mL at 37 °C.

---

Surface water is being increasingly utilized as a source of drinking water. Since here pollution can play a greater role than in ground water, the EEC published guidelines in 1975 bringing together the quality requirements for surface water (Table 32).

**Table 32:** EEC Guideline on quality requirements of surface water for drinking water preparation (normal values)

| Parameter | $A_1$ | $A_2$ | $A_3$ |
|---|---|---|---|
| color after single filtration (Pt mg/L) | 20 | 100 | 200 |
| temperature (°C) | 25 | 25 | 25 |
| ammonium (mg/L) | - | 1.5 | 4 |
| nitrate (mg/L) | 50 | 50 | 50 |
| fluoride (mg/L) | 1.5 | - | - |
| dissolved iron (mg/L) | 0.3 | 2 | - |
| copper (mg/L) | 0.05 | - | - |
| zinc (mg/L) | 3 | 5 | 5 |
| arsenic (mg/L) | 0.05 | 0.05 | 0.1 |
| cadmium (mg/L) | 0.005 | 0.005 | 0.005 |
| total chromium (mg/L) | 0.05 | 0.05 | 0.05 |
| lead (mg/L) | 0.05 | 0.05 | 0.05 |
| selenium (mg/L) | 0.01 | 0.01 | 0.01 |
| mercury (mg/L) | 0.001 | 0.001 | 0.001 |
| barium (mg/L) | 0.01 | 1 | 1 |
| cyanide (mg/L) | 0.05 | 0.05 | 0.05 |
| sulfate (mg/L) | 250 | 250 | 250 |
| dissolved or emulsified hydrocarbons (mg/L) | 0.05 | 0.2 | 1 |
| polycyclic aromatic hydrocarbons (mg/L) | 0.0002 | 0.0002 | 0.001 |
| total pesticides (mg/L) | 0.001 | 0.0025 | 0.005 |

Category $A_1$:
simple physical treatment and disinfection (e.g. rapid filtration and disinfection)
Category $A_2$:
normal physical and chemical treatment and disinfection (e.g. pre-chlorination, coagulation, flocculation, decantation, filtration, disinfection)
Category $A_3$:

intensive physical and chemical treatment (e.g. break-point-chlorination, coagulation, flocculation, decantation, filtration, adsorption on active charcoal, disinfection (ozone, chlorine) )

In Table 33, the relevant substances are listed from the draft of: Regulations governing the processing of drinking water in the Federal Republic of Germany (1987).

**Table 33:** Use of additives for drinking water treatment

| Substance | maximum permitted | maximum permitted in ready drinking water |
|---|---|---|
| chlorine, Na-, Ca-, Mg-hypochlorite chloride of lime | 1.2 - 1.8 mg/L $Cl_2$ | 0.3 - 0.6 mg/L $Cl_2$ |
| chlorine dioxide | 0.4 mg/L $ClO_2$ | 0.2 - 0.4 mg/L $ClO_2$ |
| ammonia, ammonium, chloride, -sulfate | 0.5 mg/L $NH_4$ | 0.5 mg/L $NH_4^+$ |
| ozone | | 0.05 mg/L $O_3$ |
| silver, silver chloride, -sulfate | 0.1 mg/L $Ag^+$ | 0.1 mg/L $Ag^+$ |
| mono- and poly-phosphate (Na,K,Ca) | 4.7 mg/L $PO_4^{3-}$ | 4.7 mg/L $PO_4^{3-}$ |
| sulfur dioxide, Na-, Ca-sulfite | 5 mg/L $SO_3^{2-}$ | 2 mg/L $SO_3^{2-}$ |
| hydrogen peroxide | 17 mg/L $H_2O_2$ | 5 mg/L $H_2O_2$ |
| ferric chloride, ferrous and ferric sulfate, iron chlorosulfate | - | 0.1 mg/L Fe |
| potassium permanganate | - | 0.05 mg/L Mn |
| aluminum sulfate, -chloride, hydroxychloride, hydroxysulfate, Na-aluminate, aluminum hydroxychlorosulfate | - | 0.2 mg/L Al |
| clay, activated charcoal | - | 0.5 mg/L |
| polyacrylamide | 1 mg/L | 0.005 mg/L |
| silicic acid, Na-silicate | 40 mg/L $SiO_2$ | - |
| sodium ions | | 150 mg/L $Na^+$ |
| magnesium ions | - | 50 mg/L $Mg^{2+}$ |

In addition to the listed substances, the following materials may also be used in the processing of drinking water:

calcium carbonate, partially roasted dolomite, calcium oxide, calcium hydroxide, magnesium carbonate, magnesium oxide, magnesium hydroxide, sodium carbonate, sodium hydroxide, sodium bicarbonate, sulfuric acid, hydrochloric acid.

### 7.4 Water for Use in Construction

Water destined for building use should satisfy certain requirements in order to avoid possible materials failure. If possible, a pure water (drinking or similar quality) should be used because impure or very salty water can reduce the hardness of e.g. cement or delay the setting process. The presence of mineral or humic acids or carbonic acid can retard the hardening of low calcium cements by reacting with the calcium before setting begins. Oils and fats can coat the reactive surface of the cement components thereby preventing entry of the water essential for hardening. Increased concentrations of organic substances can delay hardening in the same way. The following waters are unsuitable for use with cement:

- sea water with more than 3.5% salt
- water with more than 3.5% dissolved sulfate
- organically polluted waste water
- water with pH 4 (may possibly be neutralised before use)

The following recommended limits (Table 34) should be obeyed for water used in especially stressed concrete e.g. foundations.

Table 34: Recommended values for water in concrete

| Substance | Threshold value |
|---|---|
| free $CO_2$ | 25 mg/L |
| pH | ca. 7 |
| sulfide | not detectable |
| sulfate | 250 mg/L |
| chloride | 1 500 mg/L |
| ammonium | 100 mg/L |
| magnesium | 200 mg/L |
| potassium permanganate consumption | 25 mg/L |
| humic acids, hydrocarbons | not detectable |

The following values should be adhered to where the concrete is reinforced with iron:

| | |
|---|---|
| chloride | 100 mg/L |
| nitrate | 20 - 50 mg/L |

Hardened concrete can be attacked by contact with water. Threshold values for the assessment of such water according to DIN 4030 are listed in Table 35.

Table 35: Threshold values for assessment of waters aggressive to concrete (DIN 4030)

| Aggressive components | Weak corrosive | Strongly corrosive | Very strongly corrosive |
|---|---|---|---|
| pH | 6.5 - 5.5 | 5.5 - 4.5 | 4.5 |
| calcium aggressive $CO_2$ (marble test according to Heyer) (mg/L) | 15 - 30 | 30 - 60 | 60 |
| ammonium (mg/L) | 15 - 30 | 30 - 60 | 60 |
| magnesium (mg/L) | 100 - 300 | 300 - 1500 | 1500 |
| sulfate (mg/L) | 200 - 600 | 600 - 3000 | 3000 |

## 7.5 Water for Irrigation

Guidelines for judging the quality of water for irrigation can only be usefully applied after consideration of climate, soil, plant types and the irrigation system in use.

A general categorization of the salt contents of waters may be made as follows (Table 36):

Table 36: Categories of water salt contents

| Degree of salt content | Amount of dissolved salts (g/L) |
|---|---|
| weakly salty | - 0.15 |
| moderately salty | 0.15 - 0.5 |
| strongly salty | 0.5 - 1.5 |
| very strongly salty | 1.5 - 3.5 |

Various cations and anions can adversely affect the irrigation:

### Magnesium

High concentrations can adversely affect plant growth. According to the formula:

$$x = \frac{Mg^{2+} \cdot 100}{Ca^{2+} + Mg^{2+}} \quad \frac{n}{2} \text{ mmol/L}$$

a standard value, x, can be calculated. A value of 50 is harmful to many plants.

### Carbonate/bicarbonate
Waters containing carbonates are harmful to alkaline soils as well as limestone-containing and compact soils. However, they can be advantageous towards acidic and sandy soils.

### Chloride
A higher chloride concentration can be harmful to many cultivated plants and especially fruit trees. Tolerance limits are known for various plant types (Table 37):

**Table 37:** Chloride tolerance limits for plants

| Effect | Chloride concentration (mg/L) |
|---|---|
| weak<br>the water is suitable for nearly all plants | < 70 |
| moderate<br>suitable for Cl-tolerant plants, otherwise weak to medium damage | 70 - 140 |
| medium<br>suitable for salt resistant plants | 140 - 280 |
| strong<br>light to moderate damage caused to salt-resistant plants | > 280 |

### Boron
At low concentrations, boron is an important element for plant growth but can have toxic effects at higher concentrations. The following tolerance limits may be given (Table 38):

**Table 38:** Boron tolerance limits for plants

| Effect | Boron concentration (mg/L) |
|---|---|
| weak<br>the water is suitable for all plants | 0.3 - 1.0 |
| medium<br>suitable for boron-tolerant plants | 1.0 - 2.0 |
| strong<br>suitable for boron-resistant plants | 2.0 - 4.0 |

## 7.5 Water for Irrigation

SAR - value

The SAR (Sodium Adsorption Ratio) is often employed in assessing the suitability of irrigation waters. The calculation is carried out according to the formula in 6.2.3.6.

An important aid in assessing water is given in Figure 47.

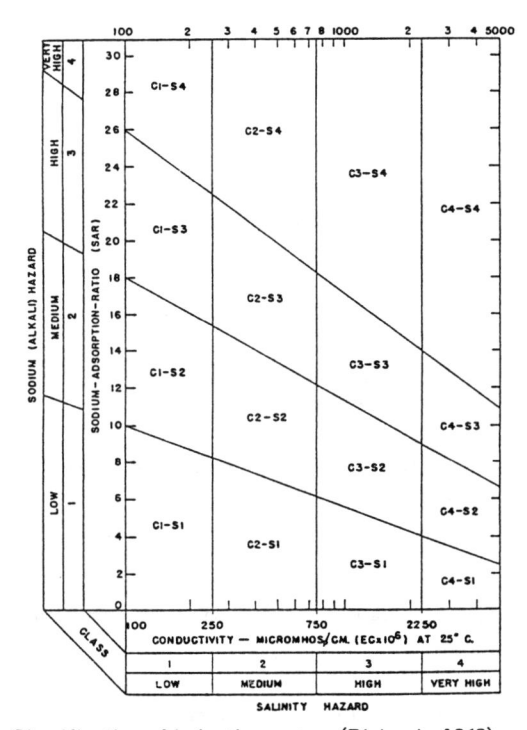

**Fig. 47:** Classification of irrigation waters (Richards, 1969)

The terms $C_1 - C_4$ and $S_1 - S_4$ have the following meaning:
- $C_1$: water with low salt content (to 0.15 g/L)
- $C_2$: water with medium salt content (0.15 - 0.5 g/L)
- $C_3$: water with high salt content (0.5 - 1.5 g/l)
- $C_4$: water with very high salt content (1.5 - 3.5 g/L)

- $S_1$: SAR < 10 in waters with low salt content
  SAR < 2.5 in waters with high salt content
- $S_2$: SAR 10 - 18 in waters with low salt content
  SAR 2.5 - 7 in waters with high salt content
- $S_3$: SAR 18 - 26 in waters with low salt content
  SAR 7 - 11 in waters with high salt content
- $S_4$: SAR > 26 in waters with low salt content
  SAR > 11 in waters with high salt content

## 7.6 Waste Water

In order to determine threshold values for the introduction of waste water, a distinction must be made between direct and indirect introduction. Direct discharges transfer treated (and sometimes untreated) waste water directly into a waterway. Indirect discharges transfer the waste water into the public drainage system and thereby usually to a sewage plant.

In general, the following protective goals should be aimed for:

| | | |
|---|---|---|
| a) | for personnel | Protection against $H_2S$, HCN, $SO_2$, $CO_2$, extreme pH, high temperatures. |
| b) | for construction | Protection against attack/damage caused by extreme pH, sulfate, aggressive $CO_2$, deposits. |
| c) | for the function of sewage plants | Protection against reduced efficiency or break-down caused by excessive contamination. |
| d) | for the water quality of the receiving water | Protection against undesired concentrations of nutrients and pollutants. |

In the direct drainage case, threshold values may be set based on the emission or immission principle.

Immission threshold values are water quality criteria which describe the condition of the water and take into account its total pollution and dynamic self-cleaning (see Table 26). This requirement is used e.g. in the EEC Guidelines on quality requirements of surface water for the preparation of drinking water (1975) (Table 32) and in the EEC Guideline on quality of bathing water (Table 28). The emission threshold values give on the other hand concentration conditions at the discharge point from the drain of a canal or sewage plant. Those rules have the advantage of being more easily monitored.

In the Federal Republic of Germany, maximum emission values have been drawn up for industrial and domestic waste water. Special conditions have in the meantime been drawn up for more than 50 industrial branches. A selection of some important areas is shown in Table 39. The threshold values are for a two hourly or 24 hourly sampling.

**Table 39:** Waste water discharge threshold values for a selection of industrial branches (2 hourly or 24 hourly sampling)

| Branch | Sediments 2 h mL/L | Sediments 24 h mL/L | COD 2 h mg/L | COD 24 h mg/L | BOD$_5$ 2 h mg/L | BOD$_5$ 24 h mg/L |
|---|---|---|---|---|---|---|
| Domestic waste water | | | | | | |
| a) 60 kg/day BOD$_5$ | 0.3 | | 180 | 120 | 45 | 30 |
| b) 60 to 600 kg/day BOD$_5$ | 0.3 | | 160 | 110 | 35 | 25 |
| c) exceeding 600 kg/day BOD$_5$ | 0.3 | | 140 | 100 | 30 | 20 |
| Sugar factories | 0.5 | | 500 | | 50 | |
| Unbleached cellulose | | 4.5 | | 120 | | 40 |
| Bleached cellulose | | 6 | | 220 | | 70 |
| Cellulose refining | | 7 | | 350 | | 120 (deciduous) 80 (coniferous) |
| Textile finishing | | | | 200 | | 30 |
| Tanneries | 0.3 | | 250 | 200 | 25 | 20 |
| Iron and steel works | 0.5 | | 100 | | | |
| Ore processing | 0.3 | | | | | |
| Coal mining | 0.3 | | 100 | | | |
| Refineries | 0.3 | | 35 | 25 | 3 | 2 |

The following guideline for industrial waste waters was drawn up in the Federal Republic of Germany by the "Abwassertechnische Vereinigung (ATV)" (Table 40). However, in most of the German cities some more stringent requirements are in use.

**Table 40:** Normal values for the indirect discharge of waste waters

| Parameter | value |
|---|---|
| Temperature | 35 °C |
| pH | 6.5 - 10 |
| Sediments | 1 mL/L after 0.5 hours |
| Saponifiable oils and fats | 250 mg/L |
| Mineral oils and fats | 20 mg/L (after prior separation) |
| Halogenated hydrocarbons | 10 mg/L |
| Arsenic | 1 mg/L |
| Lead | 2 mg/L |
| Cadmium | 0.5 mg/L |
| Chromium | 3 mg/L |
| Chromium VI | 0.5 mg/L |
| Copper | 2 mg/L |
| Nickel | 4 mg/L |

**Table 40:** continued

| Parameter | Value | |
|---|---|---|
| Silver | 0.05 | mg/L |
| Selenium | 1 | mg/L |
| Zinc | 5 | mg/L |
| Tin | 5 | mg/L |
| Ammonium | 200 | mg/L |
| Total cyanides | 20 | mg/L |
| part easily released | 1 | mg/L |
| Fluoride | 60 | mg/L |
| Nitrite | 20 | mg/L (in the case of larger loadings) |
| Sulfate | 600 | mg/L |
| Sulfide | 2 | mg/L |
| Volatile phenols | 100 | mg/L |

## 7.7 Soil

The testing of soil can provide important information on nutrient deficiency or excess. Because of the often difficult problems of assessment, especially of tropical and sub-tropical soil, no summary of relevant criteria can be given here. The reader is therefore referred to specialist literature (Literature list).

In soil having a high salt content, recommendations concerning fertilizer treatment can only be given after the whole mineral content is known. Cultivated crops in salty soil suffer not only from a lack of water but often also from nutrition problems. An alkaline salty soil usually refers to a soil whose extract (1 : 5) has an electrical conductivity exceeding 4 mS cm$^{-1}$ and whose SAR-value (section 6.2.3.6) exceeds 13. In soil rich in alkaline earths, the corresponding values are > 4 mS cm$^{-1}$ and SAR < 13.

Heavy metals are usually introduced into the soil via domestic sewage sludge. The following limits on the spreading of sewage sludge are in effect in the Federal Republic of Germany (Table 41).

**Table 41:** Heavy metal threshold values for sewage sludge spreading on agriculture land

a) Sewage sludge may only be spread on agricultural land without official permission when the following heavy metal concentrations are not exceeded:

**Table 41:** continued:

| Parameter | Threshold Value |
|---|---|
| Lead | 1 200 mg/kg |
| Cadmium | 20 mg/kg |
| Chromium | 1 200 mg/kg |
| Copper | 1 200 mg/kg |
| Nickel | 200 mg/kg |
| Mercury | 25 mg/kg |
| Zinc | 3 000 mg/kg |

b) The spreading of sewage sludge on agricultural land is forbidden when at least one of the following values is exceeded in the soil tested:

| Parameter | Threshold Value |
|---|---|
| Lead | 100 mg/kg |
| Cadmium | 3 mg/kg |
| Chromium | 100 mg/kg |
| Copper | 100 mg/kg |
| Nickel | 50 mg/kg |
| Mercury | 2 mg/kg |
| Zinc | 300 mg/kg |

The laying of pipes in the ground requires that certain soil characteristics which can affect the corrosion of iron and steel be considered. The more important parameters are as follows:

Soil type
Sandy and chalky soils and well aired loams are generally not aggressive. Peaty, chalk-free humus and muddy soils are known to be aggressive. Deposited soil (slag, refuse) is also usually aggressive.

Soil humidity
In aggressive soil the corrosiveness is greatest at a water content of approx. 20 %.

pH
In soil having a pH below 6 (suspension with distilled water), the aggressiveness increases with falling pH.

Total acidity
Soil with a total acidity (to pH 7) exceeding 25 mL of 0.1 M NaOH/kg is classified as being aggressive.

Chalk content
Aerobic soil with a calcium carbonate content exceeding 5% is not aggressive in the absence of higher sulfate concentrations.

Carbon
Soil containing elementary carbon is classified as aggressive owing to the possibility of galvanic element formation.

Chloride
Chloride concentrations (measured in aqueous extract) exceeding 100 mg/kg soil promote corrosion.

Sulfate
Sulfate ions can promote corrosion when their concentration (measured in aqueous extract) exceeds 200 mg/kg soil. In aerobic soils containing calcium carbonate more than 5 %, sulfate concentrations up to 500 mg/kg soil are harmless.

**Literature**

American Public Health Association (ed.) (1976): Standard Methods for the Examination of Water and Waste Water. APHA, Washington D.C.

ASTM (ed.) (1977): Annual Book of ASTM Standards, Part 31: Water. ASTM, Philadelphia.

Blitz E., Haug H., Pöppinghaus K., Rump. H.H. (1981): Traitment des eaux résiduaires industrielles. GTZ-Seminaire à Tunis, Oct. 1980. GTZ, Eschborn.

Brady N. (1974): The Nature and Properties of Soils. Macmillan, New York.

Cheeseman R., Wilson A. (1978): Manual on analytical quality-control for the water industry. Water Research Centre, Stevenage.

Daubner J. (1983): Mikrobiologie des Wassers. Akademie-Verlag, Berlin.

DVWK (ed.) (1983): Hydrogeologische Aspekte der Grundwasserchemie. Fortbildungslehrgang Grundwasser, Bonn.

Environmental Protection Agency (ed.) (1979): Methods for Chemical Analysis of Water and Wastes. EPA, Cincinnati.

Fachgruppe Wasserchemie in der GdCh (ed.): Deutsche Einheitsverfahren zur Wasser-, Abwasser- und Schlammuntersuchung. Verlag Chemie, Weinheim.

FAO/UNESCO (ed.) (1973): Irrigation, Drainage and Salinity. FAO, Paris.

Frevert T. (1983): Hydrochemisches Grundpraktikum. Birkhäuser, Basel.

Förstner U., Wittmann G. (1981): Metal Pollution in the Aquatic Environment. Springer, Berlin.

Funk W. et al. (ed.) (1985): Statistische Methoden in der Wasseranalytik. VCH-Verlagsgesellschaft, Weinheim.

Gradl T. (1981): Leitfaden der Gewässergüte. Oldenbourg, München.

GTZ (ed.): Methodes pour l'analyse des eaux, Vol. 1-3. GTZ, Eschborn.

GTZ (ed.) (1980): Technologie de l'eau potable. GTZ, Eschborn.

GTZ (ed.) (1984): Abwassertechnologie. Springer, Berlin.

# Literature

Herrmann R. (1977): Einführung in die Hydrologie. Teubner, Stuttgart.

Hutton L. (1983): Field Testing of Water in Developing Countries. Water Research Centre, Medmenham.

Institut für Wasserwirtschaft (Hrsg.) (1976): Ausgewählte Methoden der Wasseruntersuchung, Bd. $\underline{1}$ + $\underline{2}$. VEB G.Fischer, Jena.

Kretzschmar, R. (1979): Kulturtechnisch-bodenkundliches Praktikum. Verlag Universität Kiel, Kiel.

Merck E. (ed.): Die Untersuchung von Wasser, 11th Ed. Darmstadt.

Richards L. (ed.) (1969): Diagnosis and Improvement of Saline and Alkali Soils. U.S. Dep. of Agriculture, Washington D.C.

Rodier J. (1978): L'analyse de l'eau. Dunod, Paris.

Roth L., Weller U. (1982): Gefährliche Chemische Reaktionen. Bd. $\underline{1}$ + $\underline{2}$. Ecomed, Landsberg.

Roth L. (1979): Sicherheitsfibel Chemie. Ecomed, München.

Sanchez P. (1976): Properties and Management of Soils in the Tropics. Wiley, New York.

Schlichting E., Blume H.P. (1966): Bodenkundliches Praktikum. Parey, Hamburg.

Sontheimer H., Spindler P., Rohmann U. (1980): Wasserchemie für Ingenieure. DVGW-Forschungsstelle, Karlsruhe.

Suess M. (WHO) (ed.) (1982): Examination of Water for Pollution Control. Vol. $\underline{1-3}$. Pergamon Press, Oxford.

Stumm W., Morgan J. (1981): Aquatic Chemistry. Wiley, New York.

Weast R. (ed.) (1984): Handbook of Chemistry and Physics. The Chemical Rubber Co., Cleveland.

WHO (ed.) 1984: Guidelines for Drinking Water Quality, Vol. $\underline{1}$ Recommendations. WHO, Geneva.

Wolff W., Schwahn M. (1980): Sicherheit im Labor. Diesterweg, Frankfurt.

## Appendix

**Table 1:** Outlier - test according to Dixon

| N | probability point |
|---|---|
| 3 | 0.941 |
| 4 | 0.765 |
| 5 | 0.642 |
| 6 | 0.560 |
| 7 | 0.507 |
| 8 | 0.554 |
| 9 | 0.512 |
| 10 | 0.477 |
| 11 | 0.447 |
| 12 | 0.422 |
| 13 | 0.399 |
| 14 | 0.377 |
| 15 | 0.356 |
| 16 | 0.337 |
| 17 | 0.320 |
| 18 | 0.305 |
| 19 | 0.292 |
| 20 | 0.280 |
| 21 | 0.270 |
| 22 | 0.260 |
| 23 | 0.251 |
| 24 | 0.243 |
| 25 | 0.236 |

N number of measurements

probability points at level of significance 95%

**Table 2:** F-test to compare two standard deviations. The table gives the upper limits of significance of the F-distribution at a 95%-level of significance.
$f_1$ = degree of freedom of the nominator
$f_2$ = degrees of freedom of the denominator

**Table 3:** t-Test: The table gives the limits of significance of the student-distribution for the one-side and for the two-sided test at different levels of signifinances.

| f \ α | 0,50 | 0,20 | 0,10 | 0,05 | 0,02 | 0,01 | 0,002 | 0,001 | 0,0001 |
|---|---|---|---|---|---|---|---|---|---|
| 1 | 1,000 | 3,078 | 6,314 | 12,706 | 31,821 | 63,657 | 318,309 | 636,619 | 6366,198 |
| 2 | 0,816 | 1,886 | 2,920 | 4,303 | 6,965 | 9,925 | 22,327 | 31,598 | 99,992 |
| 3 | 0,765 | 1,638 | 2,353 | 3,182 | 4,541 | 5,841 | 10,214 | 12,924 | 28,000 |
| 4 | 0,741 | 1,533 | 2,132 | 2,776 | 3,747 | 4,604 | 7,173 | 8,610 | 15,544 |
| 5 | 0,727 | 1,476 | 2,015 | 2,571 | 3,365 | 4,032 | 5,893 | 6,869 | 11,178 |
| 6 | 0,718 | 1,440 | 1,943 | 2,447 | 3,143 | 3,707 | 5,208 | 5,959 | 9,082 |
| 7 | 0,711 | 1,415 | 1,895 | 2,365 | 2,998 | 3,499 | 4,785 | 5,408 | 7,885 |
| 8 | 0,706 | 1,397 | 1,860 | 2,306 | 2,896 | 3,355 | 4,501 | 5,041 | 7,120 |
| 9 | 0,703 | 1,383 | 1,833 | 2,262 | 2,821 | 3,250 | 4,297 | 4,781 | 6,594 |
| 10 | 0,700 | 1,372 | 1,812 | 2,228 | 2,764 | 3,169 | 4,144 | 4,587 | 6,211 |
| 11 | 0,697 | 1,363 | 1,796 | 2,201 | 2,718 | 3,106 | 4,025 | 4,437 | 5,921 |
| 12 | 0,695 | 1,356 | 1,782 | 2,179 | 2,681 | 3,055 | 3,930 | 4,318 | 5,694 |
| 13 | 0,694 | 1,350 | 1,771 | 2,160 | 2,650 | 3,012 | 3,852 | 4,221 | 5,513 |
| 14 | 0,692 | 1,345 | 1,761 | 2,145 | 2,624 | 2,977 | 3,787 | 4,140 | 5,363 |
| 15 | 0,691 | 1,341 | 1,753 | 2,131 | 2,602 | 2,947 | 3,733 | 4,073 | 5,239 |
| 16 | 0,690 | 1,337 | 1,746 | 2,120 | 2,583 | 2,921 | 3,686 | 4,015 | 5,134 |
| 17 | 0,689 | 1,333 | 1,740 | 2,110 | 2,567 | 2,898 | 3,646 | 3,965 | 5,044 |
| 18 | 0,688 | 1,330 | 1,734 | 2,101 | 2,552 | 2,878 | 3,610 | 3,922 | 4,966 |
| 19 | 0,688 | 1,328 | 1,729 | 2,093 | 2,539 | 2,861 | 3,579 | 3,883 | 4,897 |
| 20 | 0,687 | 1,325 | 1,725 | 2,086 | 2,528 | 2,845 | 3,552 | 3,850 | 4,837 |
| 21 | 0,686 | 1,323 | 1,721 | 2,080 | 2,518 | 2,831 | 3,527 | 3,819 | 4,784 |
| 22 | 0,686 | 1,321 | 1,717 | 2,074 | 2,508 | 2,819 | 3,505 | 3,792 | 4,736 |
| 23 | 0,685 | 1,319 | 1,714 | 2,069 | 2,500 | 2,807 | 3,485 | 3,767 | 4,693 |
| 24 | 0,685 | 1,318 | 1,711 | 2,064 | 2,492 | 2,797 | 3,467 | 3,745 | 4,654 |
| 25 | 0,684 | 1,316 | 1,708 | 2,060 | 2,485 | 2,787 | 3,450 | 3,725 | 4,619 |
| 26 | 0,684 | 1,315 | 1,706 | 2,056 | 2,479 | 2,779 | 3,435 | 3,707 | 4,587 |
| 27 | 0,684 | 1,314 | 1,703 | 2,052 | 2,473 | 2,771 | 3,421 | 3,690 | 4,558 |
| 28 | 0,683 | 1,313 | 1,701 | 2,048 | 2,467 | 2,763 | 3,408 | 3,674 | 4,530 |
| 29 | 0,683 | 1,311 | 1,699 | 2,045 | 2,462 | 2,756 | 3,396 | 3,659 | 4,506 |
| 30 | 0,683 | 1,310 | 1,697 | 2,042 | 2,457 | 2,750 | 3,385 | 3,646 | 4,482 |
| 35 | 0,682 | 1,306 | 1,690 | 2,030 | 2,438 | 2,724 | 3,340 | 3,591 | 4,389 |
| 40 | 0,681 | 1,303 | 1,684 | 2,021 | 2,423 | 2,704 | 3,307 | 3,551 | 4,321 |
| 45 | 0,680 | 1,301 | 1,679 | 2,014 | 2,412 | 2,690 | 3,281 | 3,520 | 4,269 |
| 50 | 0,679 | 1,299 | 1,676 | 2,009 | 2,403 | 2,678 | 3,261 | 3,496 | 4,228 |
| 60 | 0,679 | 1,296 | 1,671 | 2,000 | 2,390 | 2,660 | 3,232 | 3,460 | 4,169 |
| 70 | 0,678 | 1,294 | 1,667 | 1,994 | 2,381 | 2,648 | 3,211 | 3,435 | 4,127 |
| 80 | 0,678 | 1,292 | 1,664 | 1,990 | 2,374 | 2,639 | 3,195 | 3,416 | 4,096 |
| 90 | 0,677 | 1,291 | 1,662 | 1,987 | 2,368 | 2,632 | 3,183 | 3,402 | 4,072 |
| 100 | 0,677 | 1,290 | 1,660 | 1,984 | 2,364 | 2,626 | 3,174 | 3,390 | 4,053 |
| 120 | 0,677 | 1,289 | 1,658 | 1,980 | 2,358 | 2,617 | 3,160 | 3,373 | 4,025 |
| 200 | 0,676 | 1,286 | 1,653 | 1,972 | 2,345 | 2,601 | 3,131 | 3,340 | 3,970 |
| 500 | 0,675 | 1,283 | 1,648 | 1,965 | 2,334 | 2,586 | 3,107 | 3,310 | 3,922 |
| 1000 | 0,675 | 1,282 | 1,646 | 1,962 | 2,330 | 2,581 | 3,098 | 3,300 | 3,906 |
| ∞ | 0,675 | 1,282 | 1,645 | 1,960 | 2,326 | 2,576 | 3,090 | 3,290 | 3,891 |
| f \ α | 0,25 | 0,10 | 0,05 | 0,025 | 0,01 | 0,005 | 0,001 | 0,0005 | 0,00005 |

f  degree of freedom
α  probability of error

**Table 4:** Significance test of correlation coefficients at different probalitities of error (5%, 1%, 0.1%).

| f | 5 % | 1 % | 0,1 % |
|---|---|---|---|
| 1 | 0,9969 | | |
| 2 | 0,9500 | 0,9900 | 0,9990 |
| 3 | 0,8783 | 0,9587 | 0,9911 |
| 4 | 0,811 | 0,917 | 0,974 |
| 5 | 0,754 | 0,875 | 0,951 |
| 6 | 0,707 | 0,834 | 0,925 |
| 7 | 0,666 | 0,798 | 0,898 |
| 8 | 0,632 | 0,765 | 0,872 |
| 9 | 0,602 | 0,735 | 0,847 |
| 10 | 0,576 | 0,708 | 0,823 |
| 11 | 0,553 | 0,684 | 0,801 |
| 12 | 0,532 | 0,661 | 0,780 |
| 13 | 0,514 | 0,641 | 0,760 |
| 14 | 0,497 | 0,623 | 0,742 |
| 15 | 0,482 | 0,606 | 0,725 |
| 16 | 0,468 | 0,590 | 0,708 |
| 17 | 0,456 | 0,575 | 0,693 |
| 18 | 0,444 | 0,561 | 0,679 |
| 19 | 0,433 | 0,549 | 0,665 |
| 20 | 0,423 | 0,537 | 0,652 |
| 21 | 0,413 | 0,526 | 0,640 |
| 22 | 0,404 | 0,515 | 0,629 |
| 23 | 0,396 | 0,505 | 0,618 |
| 24 | 0,388 | 0,496 | 0,607 |
| 25 | 0,381 | 0,487 | 0,597 |
| 26 | 0,374 | 0,478 | 0,588 |
| 27 | 0,367 | 0,470 | 0,579 |
| 28 | 0,361 | 0,463 | 0,570 |
| 29 | 0,355 | 0,456 | 0,562 |
| 30 | 0,349 | 0,449 | 0,554 |
| 35 | 0,325 | 0,418 | 0,519 |
| 40 | 0,304 | 0,393 | 0,490 |
| 50 | 0,273 | 0,354 | 0,443 |
| 60 | 0,250 | 0,325 | 0,408 |
| 70 | 0,232 | 0,302 | 0,380 |
| 80 | 0,217 | 0,283 | 0,357 |
| 90 | 0,205 | 0,267 | 0,338 |
| 100 | 0,195 | 0,254 | 0,321 |
| 120 | 0,178 | 0,232 | 0,294 |
| 150 | 0,159 | 0,208 | 0,263 |
| 200 | 0,138 | 0,181 | 0,230 |
| 250 | 0,124 | 0,162 | 0,206 |
| 300 | 0,113 | 0,146 | 0,188 |
| 350 | 0,105 | 0,137 | 0,175 |
| 400 | 0,0978 | 0,128 | 0,164 |
| 500 | 0,0875 | 0,115 | 0,146 |
| 700 | 0,0740 | 0,0972 | 0,124 |
| 1000 | 0,0619 | 0,0813 | 0,104 |
| 1500 | 0,0505 | 0,0664 | 0,0847 |
| 2000 | 0,0438 | 0,0575 | 0,0734 |

**BASIC-Programs**

Prog. 1: BASIC-Program

for the calculation of mean and standard deviation

The program is self-explanatory

```
10   PRINT "MEAN AND STANDARD DEVI
     ATION"
20   PRINT
30   S = 0
40   S1 = 0
50   PRINT "NUMBER OF VALUES KNOWN
     ? (Y/N)"
60   INPUT U$
70   IF U$ = "Y" THEN 100
80   IF U$ = "N" THEN 190
90   GOTO 50
100  PRINT "NUMBER OF VALUES"
110  INPUT N
120  FOR K = 1 TO N
130  PRINT K;
140  INPUT W
150  S = S + W
160  S1 = S1 + W * W
170  NEXT K
180  GOTO 290
190  PRINT "COMPUTER NEEDS A FIGU
     RE TO TERMINATE CALCULATION.
     GIVE IN THIS FIGURE"
200  INPUT E
210  N = 0
220  PRINT N + 1;
230  INPUT W
240  IF W = E THEN 290
250  S = S + W
260  S1 = S1 + W * W
270  N = N + 1
280  GOTO 220
290  PRINT "CORRECTION ? (Y/N)"
300  INPUT U$
310  IF U$ = "Y" THEN 340
320  IF U$ = "N" THEN 430
330  GOTO 290
340  PRINT "GIVE IN THE WRONG FIG
     URE"
350  INPUT W
360  S = S - W
370  S1 = S1 - W * W
380  PRINT "GIVE IN THE RIGHT FIG
     URE"
390  INPUT W
400  S = S + W
410  S1 = S1 + W * W
420  GOTO 290
430  M = S / N
440  A =   SQR ((S1 - S * S / N) /
     (N - 1))
450  R = 100 * A / M
460  PRINT
470  PRINT "NUMBER OF VALUES
         ",N
480  PRINT "MEAN
         ",M
490  PRINT "STANDARD DEVIATION (A
     BS)",A
500  PRINT "STANDARD DEVIATION (R
     EL)",R
510  END
```

## Prog. 2: BASIC-program

### for the calculation of an anion-cation-balance

The program is self-explanatory

```
1   REM    ANION-CATION-BALANCE
10  PRINT "ANION-CATION-BALANCE"
20  PRINT
30  PRINT "1.CATIONS"
40  PRINT "PH-VALUE ?"
50  INPUT PH
60  PRINT
70  PRINT "GIVE IN CONCENTRATIONS
       OF THE FOLLOWING CATIONS IN
       MG/L"
80  PRINT
90  PRINT "SODIUM ?"
100 INPUT A
110 KA = A / 22.99
120 PRINT
130 PRINT "POTASSIUM ?"
140 INPUT B
150 KB = B / 39.1
160 PRINT
170 PRINT "AMMONIUM ?"
180 INPUT C
190 KC = C / 18.04
200 PRINT
210 PRINT "MAGNESIUM ?"
220 INPUT D
230 KD = D / 12.15
240 PRINT
250 PRINT "CALCIUM ?"
260 INPUT E
270 KE = E / 20.04
280 K = 10 ^( - PH) + KA + KB +
       KC + KD + KE
290 PRINT
300 K = K + KF
310 PRINT "ANOTHER CATION ? (Y/N
       )"
320 INPUT L$
330 IF L$ = "N" THEN  GOTO 410
340 PRINT
350 PRINT "GIVE IN CONCENTRATION
       OF THIS CATION"
360 INPUT F
370 PRINT "EQUIVALENT MASS OF TH
       IS CATION ?"
380 INPUT G
390 KF = F / G
400 GOTO 290
410 PRINT
420 PRINT
430 PRINT "SUM OF CATIONS IN MEQ
       /L =";K
440 PRINT
450 PRINT
460 PRINT "2. ANIONS"
470 PRINT "ALKALINITY BY TITRATI
       ON TO PH 4.3?"
480 INPUT M
490 PRINT
500 PRINT "GIVE IN CONCENTRATION
       IN MG/L"
510 PRINT
520 PRINT "NITRATE ?"
530 INPUT Z
540 AZ = Z / 62.01
550 PRINT
560 PRINT "CHLORIDE ?"
570 INPUT Y
580 AY = Y / 35.45
590 PRINT
600 PRINT "SULPHATE ?"
610 INPUT X
620 AX = X / 48.03
630 SA = M + AZ + AY + AX
640 PRINT
650 SA = SA + AW
660 PRINT "ANOTHER ANION ? (Y/N)
       "
670 INPUT L$
680 IF L$ = "N" THEN  GOTO 760
690 PRINT
700 PRINT "GIVE IN CONCENTRATION
       OF THIS ANION"
710 INPUT W
720 PRINT "EQUIVALENT MASS OF TH
       IS ANION ?"
730 INPUT V
740 AW = W / V
750 GOTO 640
760 PRINT
770 PRINT
780 PRINT "SUM OF CATIONS IN MEQ
       /L = ";K
790 PRINT "SUM OF ANIONS IN MEQ/
       L =";SA
800 PRINT
810 AF = ABS ((SA - K) / K * 100
       )
820 PRINT "ANALYTICAL ERROR IN P
       ERCENT = ";AF
830 END
```

Prog. 3: BASIC-program

for the calculation of regression and calibration graphs

The program is self-explanatory

```
15000  HOME
22003  DIM X9(20),Y9(20)
22007  INPUT "NUMBER OF MEASUREME
       NTS=";NQ
22010  FOR KQ = 1 TO NQ
22020  INPUT "X=";XQ(KQ)
22030  INPUT "Y=";YQ(KQ)
22040  NEXT KQ
22510  S1 = 0:S2 = 0:S4 = 0
22520  FOR KQ = 1 TO NQ
22530  S1 = S1 + XQ(KQ)
22540  S2 = S2 + YQ(KQ)
22550  S3 = S3 + XQ(KQ) * YQ(KQ)
22560  S4 = S4 + XQ(KQ) ^2
22561  S5 = S5 + YQ(KQ) ^2
22570  NEXT KQ
22580  BQ = (S3 - S1 * S2 / NQ) /
       (S4 - S1 ^2 / NQ)
22590  AQ = (S2 / NQ) - BQ * S1 /
       NQ
22591  RQ = (S3 - S1 * S2 / NQ) ^
       2 / (S4 - S1 ^2 / NQ) / (S5
       - S2 ^2 / NQ)
22600  PRINT "A=";AQ
22601  PRINT "B=";BQ
22602  PRINT "CORRELATION (R2)=;
       RQ
22605  GOSUB 22610
22606  GOTO 22605
22610  INPUT "GIVE IN MEASURED VA
       LUE";XQ
22620  B1 = BQ * XQ + AQ
22630  PRINT "Y=";B1
22635  PRINT ""
22640  PRINT "FURTHER VALUES?(Y/N
       )"
22645  GET QH$
22655  IF QH$ = "N" THEN  GOTO 15
       000
22700  IF QH$ = "J" THEN  RETURN

22710  RETURN
```

# Index

Accident prevention 1 ff.
Acidity 79 ff.
Agar culture medium 146 ff.
Alkalinity 85 ff.
Ammonia 86 ff.
Analytical goals 29 f.
- for waste water 29
- for water 29
- for soil 30
Analytical results 24 ff.
Ash residue 105 ff.

Bacillus anthracis 135
Background 157
Bathing water 159
BOD 84, 88 ff.
Boron 86, 91 f.
- tolerancy of plants 169
Burns 9

Calcium 85 f., 92 ff. 155
Calcium carbonate aggression 81 ff.
Calibration curve 20 f.
Carbon analysis
- in soil 84 f.
Carbon
- in soil 175
Catchment area 49 f.
Carbonate
- in soil 153 f., 175
- in water 169
Channels 53 f.
Check list 66
Chemicals 2 f.
- transport of 2
- storage of 2 f.
Chloride 86, 99 ff.
- tolerancy of plants 169
- in soil 175
Chlorine 75 ff. 135
- free 75 f.
- bound 75 f.

Citrobacter 142
Clostridium tetani 135
Clostridium perfringens 144
COD 84, 90, 94 ff.
Coliforms 135, 139 ff., 142 f.
Coli titer 143
Color 65
"Colored series" 140
Concrete 117, 167 f.
Concrete
- aggressiveness to 117
Concrete
- mixing water for 167
Control chart 23 f.
Correlation 18 f.
- coefficient 18, 26 f., 29
Corrosive burns 9
Corrosion 72 f., 100, 133
- in soil 174 f.
Copper 100 f.
Culture media 136, 140 f.
- preparation of 145 ff.
Cyanides 9, 101 ff.
- total 102 ff.
- easily released 104 f.

Detection limit 25
Determination level 25, 84
Dip-slide-method 138 f.
Direct discharge 171 f.
Disinfection 136 f.
Documentation 24 ff.
Drinking water 37 ff., 62, 108, 159 ff.
- processing 166
- control 38
- regulations 164

E. coli 135, 139 ff., 142 f.
EEC-regulations
- for drinking water 159 ff.
- for surface water 165
Electrical conductivity 61, 71 f. 173

# Index

Electric current 7 f.
Endoagar culture medium 147 f.
Enterobacteriaceae 135, 142, 145
Error
- systematic 10, 15
- random 10
Erwinia 135
Eutrophication 122
Evaporation residue 105 f.
Exchange capacity 153

Fecal streptococci 158
Filtreable substances 105 f.
Fire fighting 7
Fire prevention 6
First aid 8 f.
Fishery waters 174
Flash point 5
Flowing water 158
Frequency distribution 12, 15
F-Test 16, 23

Gelatine culture medium 146
Groundwater 30 ff. 57, 61 f. 157
- analysis 32 ff.
- formation of 31
- quality 31 f.
- type 32

Heavy metals
- in soil 173
H-value 151
Hydrometric propeller 54

Indirect discharge 172 f.
Ion balance 28
Iron 84 f., 108 ff., 112
- II 108 f.
- total 109 f.
Irrigation 168 ff.

Klebsiella 142
Kjeldahl nitrogen 110 ff.

Lactose-pepton culture medium 162
Magnesium 84 f., 92 ff., 155, 168 f.
Manganese 112 ff.
Matrix problems 84
Mean 12 f., 23, 26
Measurements
- local 65 ff.
- laboratory 84 ff.
Measuring weir 55 f.
Membrane filter method 138 f., 142 f., 148
Mesh size 150 f.
Moulds 145
Mycobacterium tuberculosis 135

Nitrate 86 f., 115 f.
Nitrification 88, 91, 116
Nitrite 86, 116 f.
Nitrogen
- in soil 85
Normal distribution 12, 13, 16

Odor 65
Oils and fats 117 ff.
Open channel 53
Organoleptic examination 65 ff.
Oxygen 70, 72, ff. 90
- saturation 75

Pesticides 44 f.
Phenol index 119 ff.
- without distillation 119 f.
- with distillation 120 f.
pH-value 68 f.
Phosphate 85
- total 124
- hydrolyzable 123 f.
- o-phosphate 122 f.
Phosphorus 84 f., 121 ff.
Photosynthesis 73
Pooled sample 48 f.
Potassium 124 ff.
Potassium permanganate
- consumption of 95 ff.

Proteus 135
Pseudomonas aeruginosa 144
Pumping time 61
Pumps 61, 135

Quality control 10 ff., 47
- background 11, 20 ff.
- routine 11, 23 ff.

Random sample 48, 52
Redox potential 70 f.
Regression 18 f., 20
Reliability 20
Run-off 48, 53 ff.

Salmonella 135
Salt content
- of soil 173 f.
- of water 105, 107 f., 168 f.
Sample container 56 f.
Sample conservation 60
Sample preparation 84 ff.
Sample transport 59 f.
Sampling 47 ff., 59 ff.
- dependant on flow 48 f.
- dependant on volume 48 f.
- dependant on time 48 f.
- device 56 ff.
- programs 47 ff.
- technique of 47 ff.
Sampling network 49 ff.
Saprobic index 35
SAR-value 154 ff., 170, 173
Selective culture media 148 f.
Self control 22
Self purification 35
Settleable matter 68
Sewage network 51
Sewage plant 88
Sewage sludge 173 f.
Shigella 135
Sodium 84 f., 127 f., 154 f.
Soil 41 ff. 84 ff., 173 ff.

- analysis 45 f., 84 ff.,
- kind of 174
- digestion 84 f.
- nutrients 43
- heavy metals in 44, 173 f.
- oversalting 173 f.
Standard addition 125
Standard deviation 13 f., 21, 23
Statistical methods 12 ff.
Sterilization 136
Stray value
- test according to DIXON 16
Sulfate 86, 128 ff.
- in soil 175
Surface water 35 ff., 58 f., 62, 157 ff.
- analysis 37 f.
- quality class 36 f.
- use of 37 f.
Surfactants 131 ff.
- anionic 131 f.
- non-ionic 132 f.

Taste 67
Temperature 67 f.
Time series 15, 26
Total bacterial count 134 f. 137 ff.
Trend 15
t-test 16 ff.
- one sample 17
- two sample 17
Tube well 61 f.
Turbidity 67

Undissolved substances 105 ff.

Variation coefficient 15
Velocity of flow 53 ff.
Vibrio cholerae 135
Visible depth 67

Waste 9 f.
Waste water 39 ff., 58 f., 62 f., 88 f., 171 ff.
- analysis 41

- characterization 40

Waste water
- discharge 51
- disposal 9 f.
- toxic effect 40

Water class 157 f.

Water quantity measurement 53 ff.

Well 61, 149

WHO-recommendation 160 ff.

Worm eggs 144 f.

Yersinia 135

Zinc 133 f.